synoptic exercises

for A level Geography

Gill Miller and Chris Lane

D0413024

Hodder & Stoughton

A MEMBER OF THE HODDER HEADLINE GROUP

Acknowledgements

The front cover illustration shows Procuratie Vecchie in high water, Piazza San Marco, Venice, reproduced courtesy of Sarah Quill/Bridgeman Art Library; image reproduced courtesy of Getty Images/Photodisc; Highbury, London, reproduced courtesy of © London Aerial Photo Library/Corbis; Itaipu Dam, reproduced courtesy of © Julia Waterlow; Eye Ubiquitous/Corbis.

The publishers would like to thank the following individuals, institutions and companies for permission to reproduce copyright illustrations in this book: Communications Department, Arsenal Stadium, pages 56 and 58; © Tom Bean/Corbis, page 48; © Edward Bent/Ecoscene, page 45; © Andrew Brown; Ecoscene/Corbis, page 85; © W. Perry Conway/Corbis, page 48; © Das Fotoarchiv, page 29; Adam Fricker at www.ruralnews.co.nz, page 99; Bruce Glass Photography 2001, page 35; Public Relations for Itaipu Binacional, pages 72 and 73; © Wolfgang Kaehler/Corbis, page 51 (left); Kennecott Minerals Company/Ron Plantz, page 51 (right); © Gill Miller, page 57; © Joseph Sohm, Visions of America/Corbis, page 32; Solent News and Photo Agency/Chris Balcombe Photos, page 23; © Mark L. Stephenson/Corbis, page 6; www.undiscoveredscotland.co.uk, page 90.

The publishers would also like to thank the following for permission to reproduce material in this book: 'Plan to Curb Houston's Pollution', reprinted with permission of Associated Press; 'Power cut plunges Brazil into chaos' by Alex Bellos, *Guardian*, 22 January 2002, reproduced © Alex Bellos; Highlands and Islands Enterprise for the distribution of employment in Highlands and Islands Enterprise area 1997; the *Independent* for permission to reproduce from 'President's Legacy: Clinton gift of Alaska wilderness outrages US loggers' by Andrew Gumbel from the *Independent* 13/01/2001.

Every effort has been made to trace and acknowledge ownership of copyright. The publishers will be glad to make suitable arrangements with any copyright holders whom it has not been possible to contact.

Orders: please contact Bookpoint Ltd, 130 Milton Park, Abingdon, Oxon OX14 4SB. Telephone: (44) 01235 827720. Fax: (44) 01235 400454. Lines are open from 9.00 – 6.00, Monday to Saturday, with a 24 hour message answering service. You can also order through our website www.hodderheadline.co.uk.

British Library Cataloguing in Publication Data
A catalogue record for this title is available from the British Library

ISBN 0 340 847018

First Published 2002
Impression number   10 9 8 7 6 5 4 3
Year   2008 2007 2006 2005 2004

Papers used in this book are natural, renewable and recyclable products. They are made from wood grown in sustainable forests. The logging and manufacturing processes conform to the environmental regulations of the country of origin.

Typeset by Fakenham Photosetting Ltd, Fakenham, Norfolk.
Printed in Great Britain for Hodder & Stoughton Educational, a division of Hodder Headline, 338 Euston Road, London NW1 3BH by J.W. Arrowsmith Ltd, Bristol.

Contents

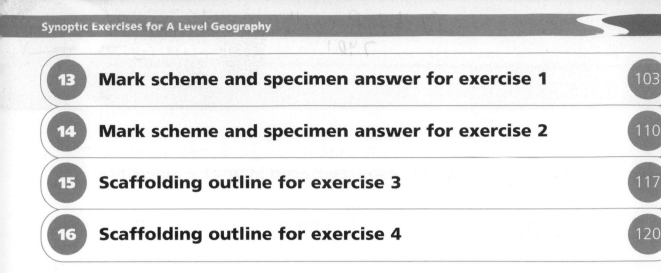

The aim of the synoptic paper is to test the geographical understanding, knowledge and skills you have acquired over the A level course. Examiners therefore expect you to be aware of the geographical concepts, processes, skills and case studies from other physical and human geography units. Don't assume, because the synoptic paper is at the end of a long line of examinations, that you can forget what has gone before – far from it. You must be tuned in to all the ideas and examples you have been taught.

Additionally, you need to be able to apply all the geographical ideas, knowledge and understanding to new situations. You should be doing this constantly throughout your course. Geography is outside your window. Open your eyes and apply what you learn in the classroom to the world outside.

Resource-based questions will contain a variety of materials to which you must respond. Practise using graphs, photographs and maps. Read sections of text and identify the key points in them. All the resources will be connected in some way and are arranged into items. Look for links between them and between the resources and your geographical studies.

The table below lists the ways in which resources may be used.

Resource skills

Photographs	Graphs
■ identification	■ annotation
■ understanding scale	■ description
■ annotation	■ analysis
■ sketching	
Maps	**Text**
■ extract information	■ identification and selectivity skills
■ identify distribution	■ classification
■ annotation	■ understanding the topic
■ sketching	■ establishes links/connections

How do you begin?

When you open the paper, read the question and have a brief look through the resources. They are presented in the order in which you need them. Identify the principal theme of the question and think back to the geography you have learned in other units.

Read the questions again and work out what resources you need for each question. The general principle is that you will build up a knowledge and understanding of the particular topic as you work through the question so that, by the end, you can develop an overview or opinion about the topic.

Read the questions carefully. The command words and mark allocation guide you towards the depth of your answer. Respond precisely to the key words, for example an 'outline' is quite brief but is more than a list; 'assess' requires that you weigh up the evidence and reach a conclusion; 'annotate' means identify in some detail a feature or trend on a resource.

Resources

Graphs

If you are asked to draw a graph, think carefully about what kind of graph is appropriate for the data you are using. It is not a test of artistic ability – graphs do not have to be perfectly presented, but they do need to illustrate the data accurately. If you are asked to draw a scatter graph, think twice before 'joining the dots'. The graph might show a trend but that does not mean that each piece of data is linked to others.

Maps

Maps are the essential tool of the geographer. They show us the patterns which we try to analyse and explain. When you first look at a map, follow the key to identify everything which is there. In particular, note the scale of the map.

Sketch maps always have a purpose. They are simple maps which identify some of the important features and relationships between them. Do not get side-tracked by interesting or quirky detail which catches your eye. Stand back from the map and look at the general pattern before you begin to work.

Photographs

Don't forget to look carefully at photographs. They, too, have a purpose in giving you a visual impression of a landscape. You should use your geographical knowledge to understand the photograph. Remember that the background in the photograph, including the sky, is as important as the foreground.

Text

When you read the sections of text, highlight the key words relevant to the question. Do not be tempted to plagiarise (i.e. copy out) sections of text in your answer. This will get you a few marks and may sound smart, but it will fall some way short of a high-level answer.

Scale

The synoptic questions may be set at any scale from global, national or regional to local scale. It is important that you appreciate the scale at which you are working. Local scale requires specific responses to particular locations, while at a national scale more general comments would be appropriate.

The question itself

Read the **Introduction** carefully. It summarises the location of the question and contains key information about its geography which you need to understand in the context of the other units you have studied. The Introduction is 'dense'. It sets the scene for the question. The aim is to tell you as much as possible about this new environment in as short a space as possible. Therefore **every word counts**.

Quality of language matters in the answers you give. Try to use appropriate geographical words; they have a precise meaning and give the impression that you know what you are talking about. Adjectives such as *undulating* or *dissected* are effective. Precision in the use of language is an important tool in the synoptic paper – an undulating landscape is not the same as a dissected one.

Data will almost always be among the resources. Look at it carefully and try to understand it. Remember that there is a difference between real numbers and percentage figures. Good students recognise this in their answers because percentage figures often mask differences in scale.

'Factors' is a word which is often used in questions. It asks you to write about the variety of things which affect a particular topic. Organise your ideas before you write and use evidence from the resources to justify your points.

'Pattern' is another commonly used word in questions. Geographers should be able to identify a pattern and try to understand it.

Look out for particular question slants that make you think. Questions may ask for **'changes in'** or **'impacts on'**. This means that you have to link the resources to your geographical understanding to appreciate what may be happening in this new situation presented in the question. Use your geographical common sense to make some sensible and relevant statements.

And, finally ... good answers are balanced, authoritative, well-reasoned and detailed.

Summary of skills for the synoptic paper

- Glance through the whole question first – the material is presented sequentially.
- There will be a variety of resources which you are expected to **use**.
- The skills section refers to material provided. Don't look for difficulties. Every word in the instruction counts (e.g. *annotate, relationship*).
- Your knowledge of the topic builds up through the question so that, by the end, you are expected to have some understanding of different aspects and draw things together.
- The final question may require integrating physical and human geographical knowledge and understanding.
- You are expected to have a good understanding of work from the synoptic links in the whole specification.
- DO NOT copy large chunks of text into your answer – unless you are desperate!

Factors

Factors tell us that you recognise the variety of things that affect a particular topic.

- Organise the factors.
- Use evidence from resources.
- Use your geographical knowledge.
- Use your geographical common sense.

Data

- Look at data carefully.
- Try to understand it – we make it as simple as possible.
- Remember the difference between real numbers and percentages.

Look out for

'changes in . . .'

'impacts on . . .'

'pattern'

Be prepared for

maps　　　graphs　　　text　　　photographs

a Explain how the physical environment poses problems for the city of Venice. **10 marks**

b Suggest how the physical environment of Venice may change
i) in the short term
ii) in the long term **12 marks**

c i) Using the data in Item 1, draw a graph to show population change in the Venice region. **6 marks**

ii) Suggest reasons for the pattern shown on the graph. **6 marks**

d To what extent are the problems facing Venice a result of the tourist industry? **8 marks**

e Identify possible strategies by which the civic authorities can ensure a sustainable future for Venice. **8 marks**

Total 50 marks

Introduction

The city of Venice lies in a shallow lagoon at the northern end of the Adriatic Sea. Venice is built on a large sand bar covering 120 closely knit islands linked by 160 canals, concentrated in an area 50km by 15km. It is linked to the mainland by rail and road along a causeway.

Each year, 7 million tourists visit Venice to enjoy unique architecture and works of art, but the local population is rapidly declining.

Venice is a car-free zone; many tourists use gondolas to travel around and most of the freight and passenger traffic is carried on the dense canal network by motor launches.

Environmental problems, such as lack of space, water pollution and the impact of global rise in sea level, are threatening the city's existence, particularly since the severe floods of 1966.

Porto Marghera on the mainland is Italy's fifth largest port, handling 27 million tonnes of cargo per year. Around 440 ships enter the lagoon annually, plus fishing vessels and passenger ships. Venice is an important stop on Mediterranean cruises, bringing 500 000 people each year.

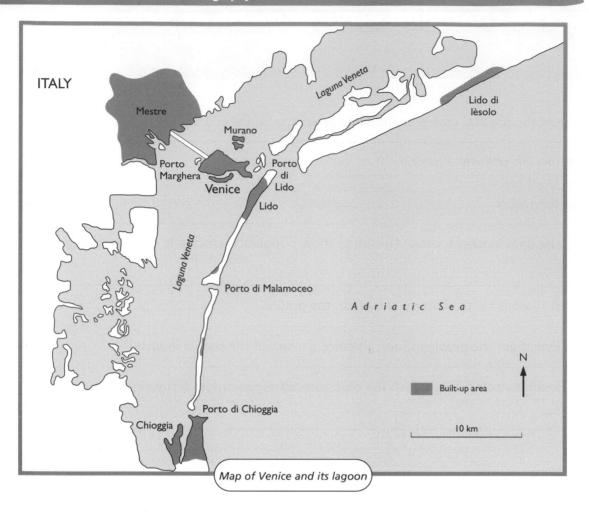

Map of Venice and its lagoon

Typical Venice scene

ITEM 1
Physical environment of Venice and its lagoon

a) Five types of environment in Venice

Lagoon	Only 1–2m deep. 'Living lagoon' in the north where a mix of freshwater and saltwater purifies and oxygenates the water. 'Dead lagoon' in the south, particularly affected by pollution.
Islands	Not covered at highest tides. Comprise 25 per cent total area.
Salt marsh vegetation (barene)	Only submerged at high tide, mainly sandbanks.
Mud flats (velme)	Submerged at every tide, covered with seaweed.
Water channels	Both natural and artificial. For centuries humans have built channels to improve communications between Venice's islands.

b) Impact of tides

Every six hours the high tide reaches the three outlets to the lagoon at the same time. This causes particular damage in artificial channels where there are no obstacles to slow down the inflow of saltwater, and fast-flowing water creates strong currents. This causes much damage to building foundations.

c) Pollution in the lagoon

Water pollution is slowly transforming the lagoon ecosystem as plant and animal species disappear. Venice has no systematic sewage treatment system and there is serious eutrophication, especially in summer. The lagoon receives waste water discharge from nearby urban centres (1 400 000 people), fertiliser run-off from mainland farms, as well as uncontrolled waste from industry on the mainland, particularly effluent from chemical, steel and oil refineries at Porto Marghera. Purification plants now treat 80 per cent of industrial waste but there are 17 abandoned rubbish dumps containing 5 million m³ of material which is slowly being washed away by rainwater and tides. The warm climate and shallow water increases risks to aquatic species and the food chain.

Sources of nutrients released into the lagoon

	Agriculture and livestock	Industries	Housing, offices, schools	Other urban land uses
Nitrogen (7000 t/year)	53%	8%	31%	8%
Phosphorous (800 t/year)	46%	8%	39%	7%

d) Consequences of pollution

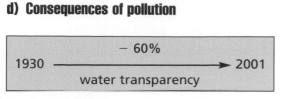

1930 ——— − 60% ——→ 2001
water transparency

Decline in transparency caused by increase in waste water in the lagoon.

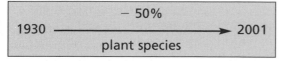

1930 ——— − 50% ——→ 2001
plant species

Coverage of lagoon bed reduced from 50 per cent to 5 per cent.

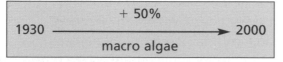

1930 ——— + 50% ——→ 2000
macro algae

Eutrophication as macro algae use available oxygen and produce sulphurous acid.

1930 ——— − 80% ——→ 2001
marine eelgrass

Eel grass should provide habitat and food for fish. Helps transform nutrients and solidifies lagoon beds.

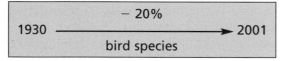

1930 ——— − 20% ——→ 2001
bird species

Reduction in diversity of badgers, otters, water fowl and small fish.

e) Flooding in Venice

Venice floods between 40 and 60 times a year as a result of high tide surges. Land reclamation has reduced the area and volume of the lagoon and hence its capacity for holding high tides. The entrances to the lagoon have been substantially deepened to allow ships to reach Porto Marghera, but this increases the ease with which high waters can reach Venice. Rivers are not replacing the lagoon sediment which is lost to sea, and this in turn increases erosion of the barene.

The pumping of groundwater between the 1930s and 1970s is causing subsidence of the city by 3 to 4cm a year. It is estimated that combined with eustatic changes, a sea level rise of only 30cm would leave St Mark's Square flooded more than 360 times each year by 2075.

ITEM 2
Environmental issues in Venice

- Wash from boats removes subsoil and erodes foundation of buildings.
- Requirement for lightweight buildings not enforced. Old, wooden, sunken piles cannot support weight of buildings which are modernised internally. Gradual subsidence.
- Exploitation of underground aquifers and natural gas encourages subsidence.
- Building facades eroded by high tides. Salt penetrates porous brickwork.

- Toxic gases, especially SO_2 from Porto Marghera, discolours buildings and causes stonework to disintegrate.
- Fungi, seaweed and lichens grow quickly in humid environment. Growth speeded up by increased CO_2 in atmosphere, linked to large-scale deforestation on the mainland.
- Buildings deteriorate when left empty as people leave Venice, or houses are bought as holiday homes.

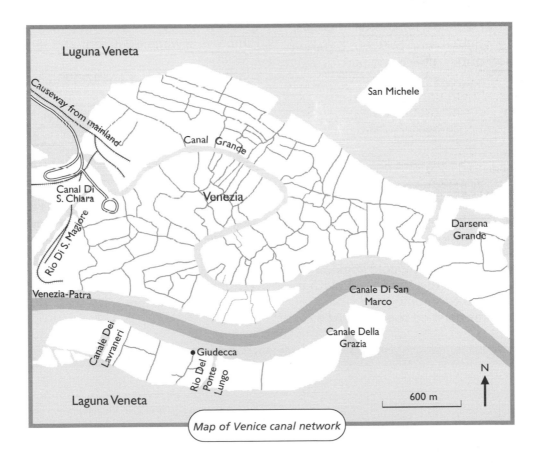

Map of Venice canal network

ITEM 3
Problems and solutions for Venice

a)

Problems	Solutions
Land subsidence	Extraction of natural gas and groundwater now prohibited
Air pollution	
Flooding	Reinforce the velme, barene and islands
Building subsidence	Reduce speed limits on waterways to reduce erosion of banks
Lack of space for new, low-cost housing	
Lack of space for industry	Build breakwaters, dykes and replenish beaches to reduce coastal erosion
Heating problems in homes	Reroute electricity supplies to protect from floodwater
Falling birth rates; fewer young people	
Tourists outnumber residents	Unload oil tankers at Trieste and transfer oil by pipeline to Porto Marghera
Purchase of second homes	Improve sewage treatment and sanitation facilities
Decline in demand for consumer goods	
Lack of co-ordination between different civic authorities, police and port	Increase number of low-cost houses for young families
	Improve quality of public services Revitalise university facilities

b) The mobile gates proposal

In December 1998, the Environment Minister decided not to approve a proposal to protect Venice from flooding using a system of mobile gates at the entrance to the lagoon. The gates would work on the same principle as the Thames barrier but the disruption to shipping and visual intrusion was deemed too great. At a cost of $6000 million compared to annual flood damage of $330 million, the gates would take money from other physical and economic developments also badly needed in Venice.

The preferred plan now is for small-scale works to protect low-lying areas within Venice.

ITEM 4

Population change in the Venice region

Population (in thousands)

	Venice city	Islands	Mainland	Total
1960	145	49	152	346
1965	124	51	189	364
1970	111	51	205	367
1975	104	50	210	364
1980	95	49	209	353
1985	86	48	200	334
1990	78	47	192	317
1995	71	45	182	298
2000	66	32	176	274

ITEM 5

Venice: the city that's loved to death

Tourists arrive in Venice on bus tours of northern Italy, sometimes 100 000 a day. As day visitors, they spend little money on hotels but frequent the fast food places and pizzerias which have grown in number. As well as adding to the overcrowding and litter, tourists contribute to the destruction of many famous monuments. Millions of hands touching the little statues in Venice have smoothed them down so that some of the features are unrecognisable.

But soon Venice is to have a make-over. Decrepit warehouses have been demolished along the Guidecca canal and a half-mile stretch of harbours, residences and customs houses is to be flattened to create space for ultra-modern designs to take Venice into the 21st century. Venetian mayor Paolo Costa said the city has to overcome its fears of modern architecture if it wants to remain vibrant and alive as opposed to a giant museum showpiece.

Water resource management in England and Wales

a i) Use the outline map (Item 1a) of the Environment Agency regions of England and Wales to represent the data in Item 1b for household water consumption. **6 marks**

ii) Using the resources in Item 1 to help you, describe and suggest reasons for the distribution you have mapped. **4 marks**

b i) Use a second copy of outline map Item 1a to represent data in Item 1b for water abstraction. **6 marks**

ii) How is the pattern of water abstraction affected by the physical environment? **6 marks**

c Suggest how water demand appears to be influenced by changes in
 i) land use **6 marks**

ii) population distribution **4 marks**

iii) changes in local economies **6 marks**

d In what ways might the changing patterns of population, industry and land use have an impact on water management in England and Wales? **12 marks**

Total 50 marks

Introduction

Water resource managers in England and Wales need to balance the increasing demand for water with available supplies. Water authorities utilise groundwater, river water and reservoirs to provide for the needs of domestic and industrial clients. Recent rapid expansion in demand, associated with changing lifestyles, increased leisure demands and the requirements of industry have resulted in the adoption of a range of more radical, less traditional, strategies. The range of water management options now includes water transfer schemes, recycling, metering and conservation measures.

Global warming suggests that the UK will face more extremes of weather conditions – more severe storms, high rainfall and more frequent drought. This adds a further dimension to the management of water across the country.

ITEM 1

Water consumption in England and Wales

a) Environment Agency regions

Outline map

b) Water abstractions and household water consumption 1998 (Environment Agency regions)

	Water abstractions 1998 (megalitres/day)	Household water consumption 1998 (litres/head/day)
North East	530	147
North West	304	138
Anglian	1032	150
Severn Trent	1072	138
Thames	1821	151
Southern	1279	158
South Western	489	156
Welsh	127	144

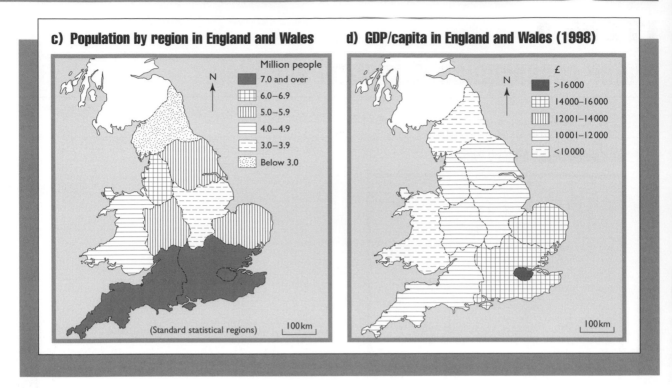

c) Population by region in England and Wales

Million people
- 7.0 and over
- 6.0–6.9
- 5.0–5.9
- 4.0–4.9
- 3.0–3.9
- Below 3.0

(Standard statistical regions)   100km

d) GDP/capita in England and Wales (1998)

£
- >16000
- 14000–16000
- 12001–14000
- 10001–12000
- <10000

100km

ITEM 2
Rainfall in England and Wales

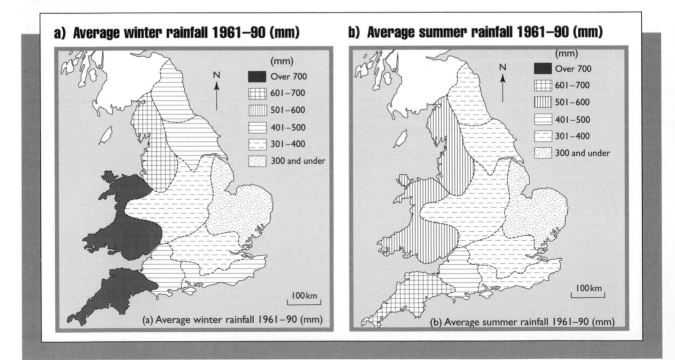

a) Average winter rainfall 1961–90 (mm)

(mm)
- Over 700
- 601–700
- 501–600
- 401–500
- 301–400
- 300 and under

(a) Average winter rainfall 1961–90 (mm)   100km

b) Average summer rainfall 1961–90 (mm)

(mm)
- Over 700
- 601–700
- 501–600
- 401–500
- 301–400
- 300 and under

(b) Average summer rainfall 1961–90 (mm)   100km

ITEM 3
Land use changes in England and Wales

a) Private house building in regions of England and Wales

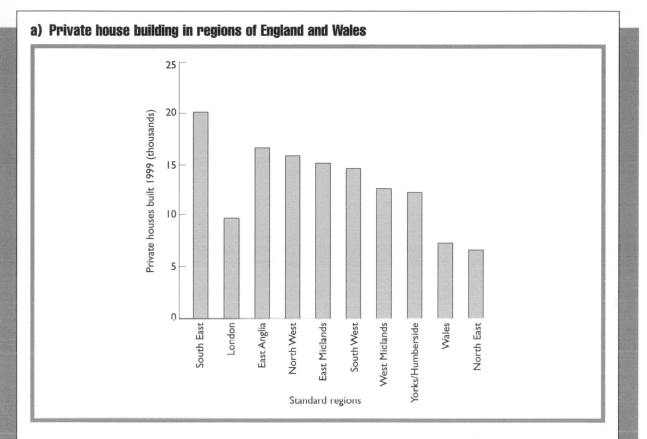

b) Land changing to urban use in England (1994)

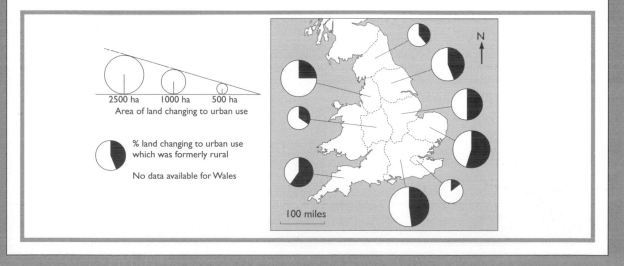

c) Inter-regional population movements in England and Wales (1998)

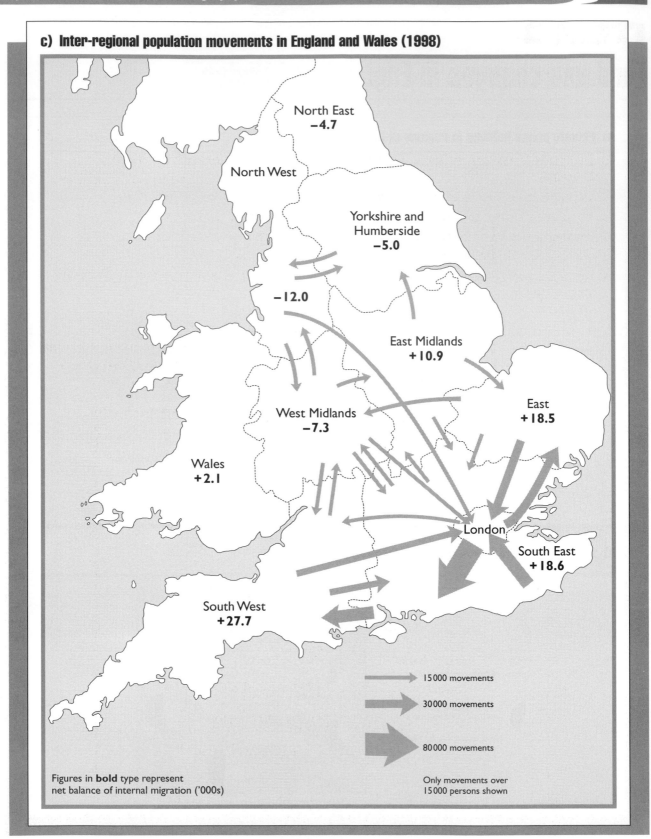

North East
−4.7

North West

Yorkshire and Humberside
−5.0

−12.0

East Midlands
+10.9

East
+18.5

West Midlands
−7.3

Wales
+2.1

London

South East
+18.6

South West
+27.7

15000 movements

30000 movements

80000 movements

Figures in **bold** type represent
net balance of internal migration ('000s)

Only movements over
15000 persons shown

ITEM 4
Demand for water

a) Influences on demand for public water supplies

- population growth and household size
- rising standards of living
- increased use of household appliances
- use of water in homes and gardens
- losses through leakage from pipes in streets and homes due to old and decaying infrastructure

b) Commercial demands for water

- intensification of agriculture, especially farmers with spray irrigation
- mineral extraction
- golf courses
- fish farms
- declining demand from manufacturing industry
- modern industry uses much less water than traditional heavy industry
- high levels of manufacturing investment in traditional industrial regions are changing the balance of water consumption

c) Manufacturing investment in England and Wales, 1994–7

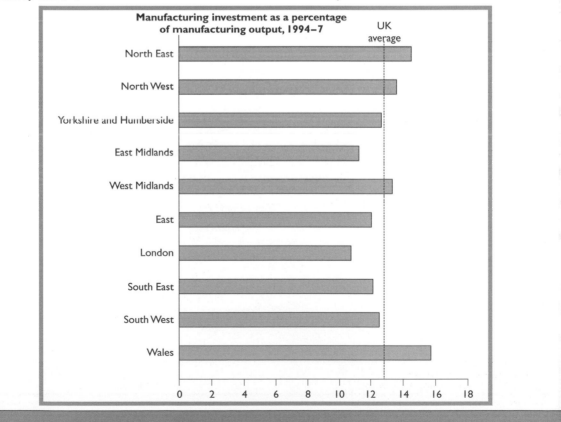

Manufacturing investment as a percentage of manufacturing output, 1994–7

d) Water management issues

- climatic change
- managing floods and drought – compromise between protecting the environment and providing for extreme weather conditions
- variations in regional development
- growth of tourism and recreation
- increasing efficiency of water recycling technologies

e) NRA water management strategy

- Reservoirs in the North and West are to be upgraded and enlarged.
- Inter-basin water transfers will deliver additional water to rivers, reservoirs and aquifers in the South and East.
- Introduction of metering and price tariffs will help to control demand and encourage consumers to use water more efficiently.

ITEM 5
Water transfer schemes in England and Wales

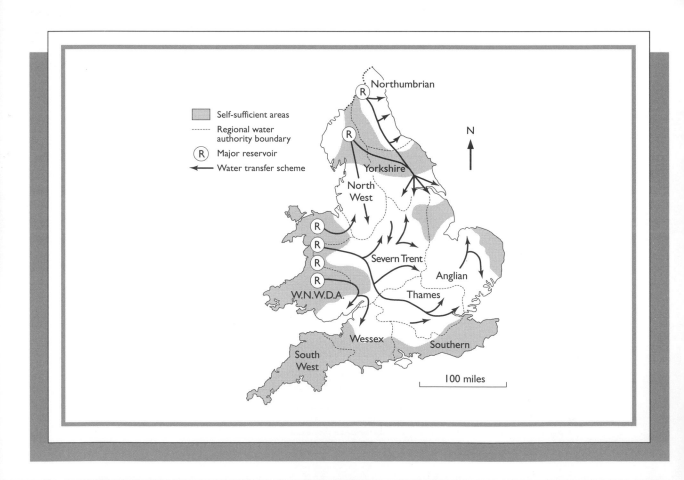

a   Outline the geographical characteristics needed for a successful container port.   **8 marks**

---

b   i)   Using Item 1a, draw a graph to compare the percentage change of container movements in the five largest ports in the UK from 1989 to 1999.   **6 marks**

---

ii)   Comment on the relative growth of container movements in these ports.   **8 marks**

---

iii)   Suggest why new developments are planned at all the major container ports in the UK.   **6 marks**

---

c   What are the advantages of Dibden Bay for a new container port?   **6 marks**

---

d   The opponents of the Dibden Bay scheme make their case on ecological, economic and environmental grounds. Assess the validity of their arguments.   **16 marks**

Total 50 marks

Introduction

In the last few years, UK container ports have experienced rapid growth as a result of general economic growth and, in particular, growth of transhipment traffic. The increasing size of container ships means that only a few British ports will be capable of handling the new generation of ships. By 2010 there is predicted to be a significant shortage of capacity for deep-sea container cargo. The major UK deep-sea container ports are Felixstowe, Southampton, Thamesport, Tilbury and Liverpool. Other ports handle short-sea or coastal transport.

In order to meet future growth in container traffic, existing ports must raise productivity and capacity. There are also a number of proposed or planned developments in the UK.

The viability of new developments depends on several factors:

- The degree to which the port can handle ships carrying over 7000 TEU vessels. (1 TEU is a measure of one 20-foot container)
- The depth of water (draft) in the port channels at low tide, and the amount of dredging necessary. Given the need to keep to tight schedules, container ships cannot be expected to wait for tides.
- The length of available quayside and the number of rows of containers has implications for the outreach of cranes.
- The amount of terminal space required to stack containers for loading and unloading.

A controversial new part is planned for Southampton Water. Opinion is sharply divided over the desirability and viability of the scheme.

Proponents

Associated British Ports
Southampton City Council
CBI (Confederation of British Industry)
TGWU (Transport & General
   Workers Union)

Opponents

RSPB
Friends of the Earth
English Nature
Council for National Parks
Hampshire County Council
New Forest District Council
Local Parish Councils
Residents Against Dibden Bay Port

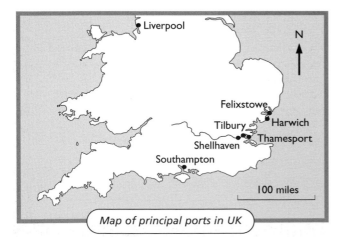

Map of principal ports in UK

ITEM 1
Container ports in the UK

a) Container movements at the largest ports

	1989		1994		1999	
	'000 units	% of all ports	'000 units	% of all ports	'000 units	% of all ports
Felixstowe	972	35.2	1231	36.6	1826	40.9
Southampton	218	7.9	411	12.2	601	13.5
Tilbury	301	10.9	344	10.2	467	10.5
Thamesport	9	0.3	161	4.8	320	7.2
Liverpool	94	3.4	257	7.6	325	7.3

b) Comparison between UK ports and European ports, 2000

	Container units per annum per metre of quay	Container units per annum per hectare of quayside
Felixstowe	971	17 880
Southampton	663	14 520
Thamesport	772	20 920
Antwerp	412	8 920
Bremerhaven	604	10 840
Hamburg	622	13 390
Le Havre	252	6 950
Rotterdam	884	16 600

c) Future developments at UK ports

- P & O Ports intends to redevelop Shellhaven refinery into the UK's largest container port.
- Felixstowe plans to use spare capacity and build one new berth.
- Tilbury is currently adding one new berth.
- Thamesport can double its capacity and develop adjacent redundant land.
- Southampton has spare capacity.
- Harwich intends to build a four-berth container port.

d) Shellhaven

The largest container port in the UK is being built at Shellhaven, the former site of Shell oil refinery, to handle 3.5 million TEU containers each year. Berths for 10 ships will be provided along 3000m of quayside. Port facilities will create 3000 jobs with 7000 more in the new transport infrastructure and light industry around the port.

ITEM 2
The Dibden Bay port proposal

Dibden Bay is located on the west bank of Southampton Water and on the edge of the New Forest. Associated British Ports (ABP) wishes to build a new terminal able to dock six large super-containers and double Southampton's container trade to 2 million tonnes of freight a year. Without it, the UK will lose out to continental ports such as Rotterdam, Hamburg and Antwerp.

Oil refineries, container terminals, marinas and urban area cover much of Southampton estuary, but the planned site is a designated SSSI and lies within the Solent and Southampton Water special protection area.

a)

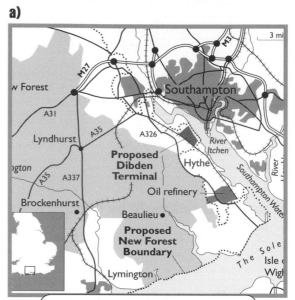

Map of proposed Dibden Terminal

b) ABP plan of proposed Dibden Bay terminal

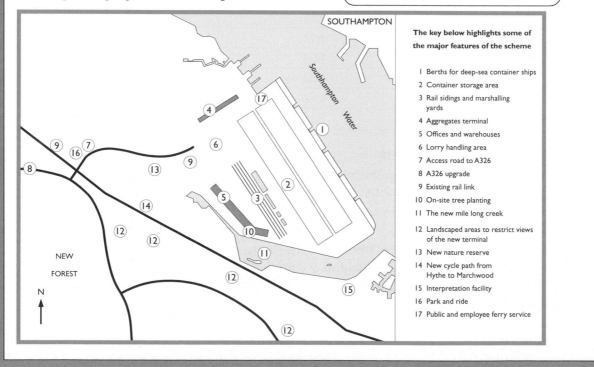

The key below highlights some of the major features of the scheme

1 Berths for deep-sea container ships
2 Container storage area
3 Rail sidings and marshalling yards
4 Aggregates terminal
5 Offices and warehouses
6 Lorry handling area
7 Access road to A326
8 A326 upgrade
9 Existing rail link
10 On-site tree planting
11 The new mile long creek
12 Landscaped areas to restrict views of the new terminal
13 New nature reserve
14 New cycle path from Hythe to Marchwood
15 Interpretation facility
16 Park and ride
17 Public and employee ferry service

The plan:

- 1850m of shipping berths and a 200ha storage/handling area
- 22 container handling cranes
- rail terminal
- public ferry service to Southampton linked to 500-space park-and-ride facility
- new road network
- creation of a tidal creek for wading birds and other wildlife, 1500m long around the southern and western edges of the terminal
- creation of 137ha wetlands, drier grasslands and woodland
- extensive landscaping of dock area – hills and 280 000 trees
- creation of 4ha of open space
- creation of a new mudflat south of Hythe to Cadland foreshore using inter-tidal sediments from the Dibden foreshore excavations, to replace the excavated area.

c) The Dibden Bay site

In the 1940s, the land was originally reclaimed with the long-term intention of port use. As Southampton port was dredged to keep it clear, a polder was created which was sealed off with an embankment in the 1960s. As this polder dried out naturally, a plant succession established which now consists of open grazing marsh and mudflats. It is noted for its diversity of birds and wildlife, and the site has become an internationally important wildlife haven.

d)

Dibden Bay, nr Southampton

ITEM 3
Controversial views about Dibden Bay

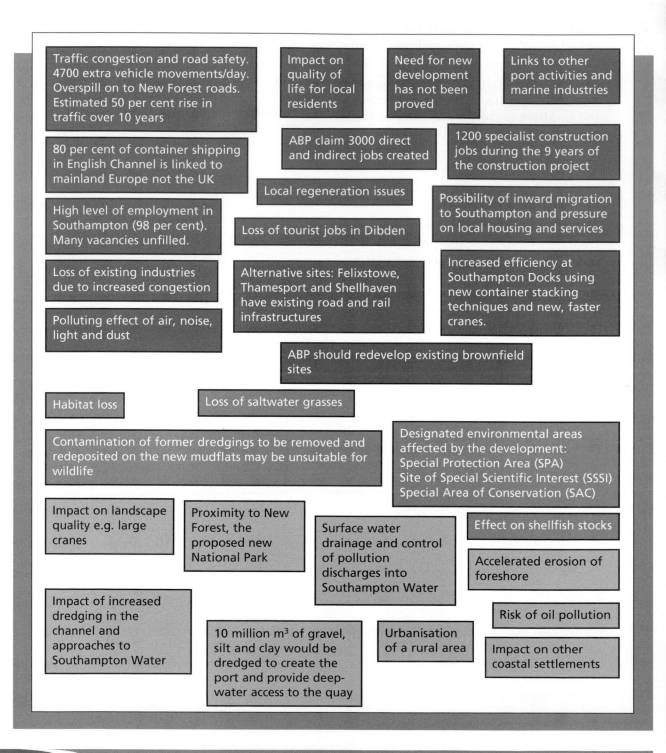

Traffic congestion and road safety. 4700 extra vehicle movements/day. Overspill on to New Forest roads. Estimated 50 per cent rise in traffic over 10 years

Impact on quality of life for local residents

Need for new development has not been proved

Links to other port activities and marine industries

80 per cent of container shipping in English Channel is linked to mainland Europe not the UK

ABP claim 3000 direct and indirect jobs created

1200 specialist construction jobs during the 9 years of the construction project

Local regeneration issues

High level of employment in Southampton (98 per cent). Many vacancies unfilled.

Loss of tourist jobs in Dibden

Possibility of inward migration to Southampton and pressure on local housing and services

Loss of existing industries due to increased congestion

Alternative sites: Felixstowe, Thamesport and Shellhaven have existing road and rail infrastructures

Increased efficiency at Southampton Docks using new container stacking techniques and new, faster cranes.

Polluting effect of air, noise, light and dust

ABP should redevelop existing brownfield sites

Habitat loss

Loss of saltwater grasses

Contamination of former dredgings to be removed and redeposited on the new mudflats may be unsuitable for wildlife

Designated environmental areas affected by the development: Special Protection Area (SPA) Site of Special Scientific Interest (SSSI) Special Area of Conservation (SAC)

Impact on landscape quality e.g. large cranes

Proximity to New Forest, the proposed new National Park

Surface water drainage and control of pollution discharges into Southampton Water

Effect on shellfish stocks

Accelerated erosion of foreshore

Impact of increased dredging in the channel and approaches to Southampton Water

10 million m³ of gravel, silt and clay would be dredged to create the port and provide deep-water access to the quay

Urbanisation of a rural area

Risk of oil pollution

Impact on other coastal settlements

a Describe and suggest reasons for the economic problems faced by the Ruhr valley in the 1980s.

12 marks

---

b Summarise the challenges to planners and developers in the Emscher valley from
i) the physical environment
ii) the social and economic environment

16 marks

---

c Assess the effectiveness of the development strategies adopted in the Emscher Valley Park.

12 marks

---

d Examine the advantages and disadvantages of large environmental management schemes such as the Emscher valley.

10 marks

Total 50 marks

Introduction

Germany has a strong global economy but, like many industrial nations, it experienced a steady decline in manufacturing employment between 1960 and 1980. The expansion of the tertiary and quaternary sectors was not sufficient to compensate for the loss of 4 million jobs in heavy industry and agriculture. This de-industrialisation was concentrated in north-western Germany, in the Ruhr region of North Rhine–Westphalia.

The Ruhr was the centre of heavy industries such as coal mining, iron and steel production and chemicals, but in the 1970s coal resources were running out, the remaining coal was expensive to mine, and the region's industries became less competitive in the global market.

As heavy industries closed down, large areas of brownfield sites were in need of substantial restoration. In 1989, the Emscher Park 10-year plan was begun to restore one of the most degraded landscapes in Europe.

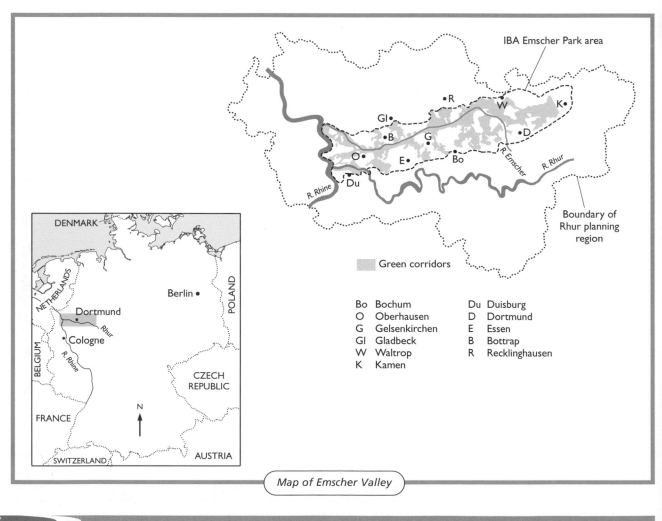

Map of Emscher Valley

ITEM 1
Economic patterns in the Ruhr

a) Urban problems of the Ruhr region

- Extensive mining created surface subsidence.
- The Emscher River was canalised in a concrete channel and became an open sewer carrying industrial and human waste.
- In the flat landscape the slag heaps were like small mountains and industrial buildings rose above 10 storeys.
- In the late 1980s, 15 per cent of the 2 million people living in the region were unemployed.

b) Employment structure in the Ruhr region

	Percentage in primary sector	Percentage in secondary sector	Percentage in tertiary sector
1960	5	62	33
1970	3	58	39
1980	2	53	45
1993	2	44	54

c) German steel production by region (million tonnes)

	Ruhr	Saar	South	Rest of Germany
1960	24.7	3.8	1.6	4.0
1970	30.5	5.4	1.9	7.2
1976	26.1	4.7	2.2	7.4

d) Unemployment and GDP per capita in the Ruhr

City in the Ruhr region	Average unemployment rate 1986–8 (average for Germany = 6.5%)	Index of GDP per capita 1986 (average for Germany = 100)
Duisburg	11.7	105.4
Essen	11.1	125.7
Oberhausen	14.7	87.8
Bottrop	10.3	66.4
Gelsenkirchen	11.6	136.6
Dortmund	12.8	101.4
Bochum	10.8	114.8

ITEM 2
The Emscher Park project

a)

The Emscher River basin is several kilometres wide and 70km long between Dortmund and Duisburg, with a population of 3 million. Cast iron foundries, coking plants, coal mines and chemical plants produced high levels of pollution, soil contamination, abandoned factories and huge smelters which disfigured the region's landscape. Emscher Park is brownfield redevelopment on a massive scale.

The North Rhine–Westphalia regional authority recognised the vast scale of rehabilitation needed. Using the land for high-tech businesses would produce immediate results for employment but the scale of the dereliction required dramatic change and a new approach to urban redevelopment. The International Building Exhibition (IBA), a comprehensive 10-year regional planning scheme, was created to help develop new and sustainable ideas for the reuse of the massive industrial wastelands, and to create a network of regional open space linking the former industrial sites.

The Emscher Park Planning Company co-ordinates individual redevelopment projects and seven 'green corridors' which form a complete park system.

b) Main principles of the redevelopment scheme

- Re-use of land to prevent additional use of natural areas or further use of under-developed areas. This also reduces costs for new infrastructure for roads and sewers.
- Maintenance, modernisation and re-use of existing buildings to extend their life.
- Environmentally friendly building practices for new buildings and rehabilitation.
- Transformation of industry towards environmentally friendly production.
- Incorporation of the industrial heritage into a cultural centre.
- An emphasis on architecture whereby building and site design are seen as key elements for a successful economic, environmental and social strategy.
- Strong emphasis on energy efficiency, especially solar power.

c) Issues concerning the Emscher project

- Job creation not the main focus.
- Effective co-ordination was vital.
- Scale of investment
- Ten-year lifetime of the IBA.
- Concentration of investment in one flagship scheme.
- Historical image of the Ruhr region may not attract modern industry.

ITEM 3
Projects in the Emscher Valley Park

Industry goes green

A water park was built as part of the ecological regeneration of the Emscher Canal system. Promenades and parks along railway lines link adjoining cities. Recreation areas are incorporated into parks.

Industry meets culture

In Emscher Park, concerts are given in former steel plants and visitors wander around waste heaps. The 12-storey Oberhausen gasometer no longer stores natural gas but is used as a space for cultural events.

Industry meets leisure

The Duisburg steel plant, closed in 1985, has been transformed into a landscape park where visitors can climb giant blast furnaces, slide down chutes or try climbing walls. Its three imposing furnaces create an impressive 'industrial-monument' skyline over what is now a huge leisure area with hiking trails.

Industry meets art

The former Zollverein colliery in Essen was once called the 'Cathedral of Labour'. It is now home to a range of theatre groups and design studios. Graphic artists and industrial designers develop innovative ideas and work particularly with small local companies.

Industry meets housing

In garden villages 6000 houses have been established, mainly built on brownfield sites. In Prosper III, near the centre of Bottrop, small businesses, commercial and retail uses are mixed with single-person homes, social housing for recent immigrants and housing for the elderly.

Industry at the harbour

Duisburg's inner harbour project seeks to transform the historic mill and grainstore buildings into a mixed-use neighbourhood and waterfront park. Low-income housing, studios, galleries and cafés plus offices and hotels form the basis for innovative design and modern planning.

Industry and the Emscher waterway

Mining and domestic waste used to flow through open concrete sewers because underground mining made it impossible to lay underground pipes. Now the whole basin is being re-engineered to restore stream channels and improve water quality, although this will take up to 50 years to complete.

Duisberg Canal, Germany

ITEM 4
Funding the Emscher valley project

a) Funding the Emscher valley project

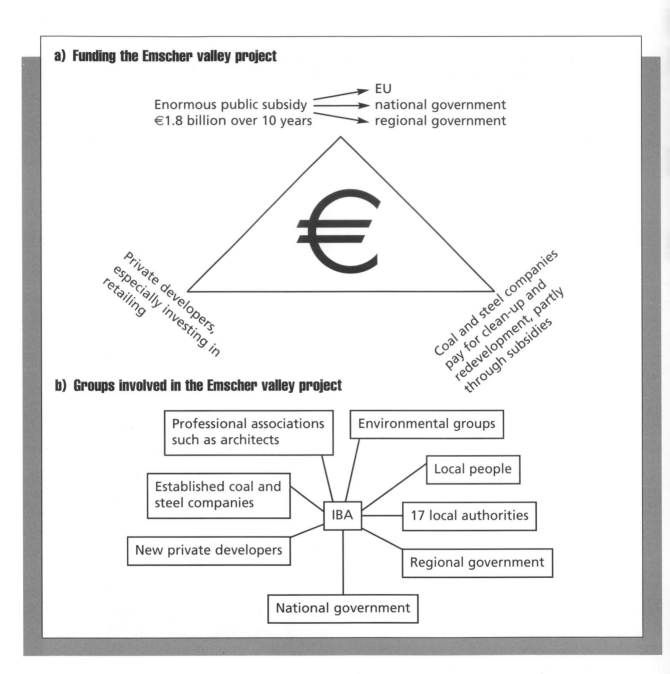

EU
Enormous public subsidy ——→ national government
€1.8 billion over 10 years ——→ regional government

Private developers, especially investing in retailing

Coal and steel companies pay for clean-up and redevelopment, partly through subsidies

b) Groups involved in the Emscher valley project

Professional associations such as architects

Environmental groups

Established coal and steel companies

Local people

IBA

17 local authorities

New private developers

Regional government

National government

a In what ways was Houston an unlikely site for extensive development?   **8 marks**

b i)   Using the data in Item 1a), complete the choropleth map of population change in Houston 1990–2000 (Item 1b).   **6 marks**

   ii) Describe and suggest reasons for the pattern shown on the map.   **10 marks**

c In what ways does the growth of Houston depend on
   i)   physical factors?
   ii) human factors?   **14 marks**

d Assess the importance of physical and human factors in creating environmental problems for Houston.   **12 marks**

Total 50 marks

Introduction

Houston (pop. 4 million) is a large city and commercial centre situated in the flat coastal plain of south-east Texas, USA, only 50 miles from the Gulf of Mexico. It sprawls over 22 500 km², larger than any other American city, and bigger even than Israel or El Salvador.

The Port of Houston exists due to the dredging of the Buffalo Bayou channel to create the Houston Ship Canal which supports oil refining and petrochemical industries. Businesses of all kinds are booming in Houston with its extensive freeway system and diverse population. It thrives because it has cheap land, immigrant labour, low taxes and pro-business policies. However, as a victim of its own success, Houston has to resolve a number of environmental issues in order to sustain future growth.

Houston – 'Bayou City'
Houston has 10 winding waterways which irrigate the lush landscape and, with the lakes and bays, provide important areas of natural habitat. There are 22 watersheds and thousands of miles of watercourses.

Annually, 127cm of rain falls, often in heavy thunderstorms, producing irregular and significant flooding. An extensive bayou system attempts to drain floodwaters from roads and neighbourhoods. However, groundwater levels remain high and new houses have sloping driveways as builders raise ground-floor levels higher and higher above the potential flood risk.

Houston city centre

ITEM 1
The growth of Houston

a) Population change for selected neighbourhoods in Houston, 1990–2000

Number	Neighbourhood name	Population change 1990–2000, + / − per cent
1	Central Southwest	24.3
2	Alief	18.1
3	Elderidge	50.8
4	Addicks Park Ten	217.8
5	Northside	19.8
6	East Houston	11.5
7	El Dorado	−10.6
8	Port Houston	−7.8
9	Clinton Park	−19.4
10	Meadowbrook	22.0
11	Greater Hobby	15.2
12	Ellington South Belt	18.2

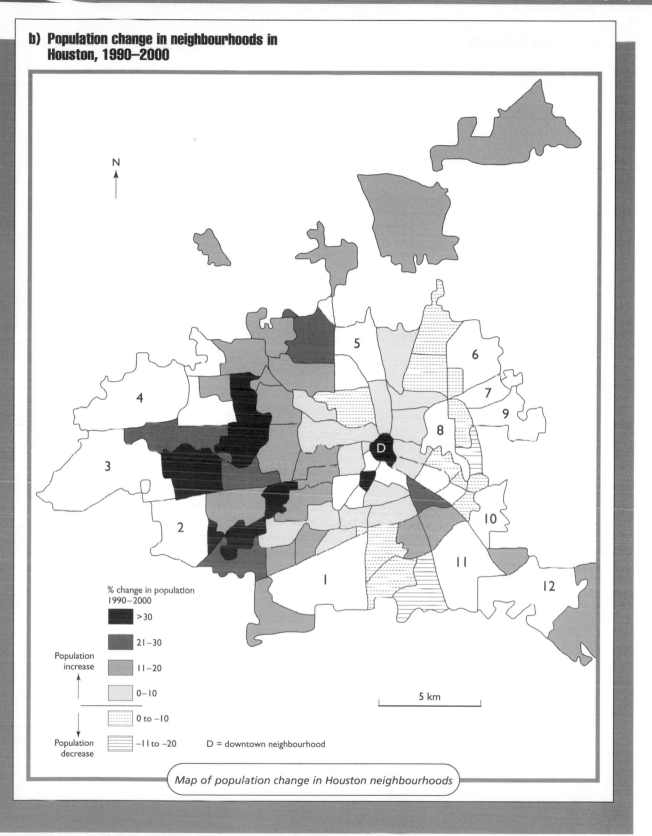

b) Population change in neighbourhoods in Houston, 1990–2000

% change in population 1990–2000

- >30
- 21–30

Population increase

- 11–20
- 0–10

Population decrease

- 0 to –10
- –11 to –20

D = downtown neighbourhood

5 km

Map of population change in Houston neighbourhoods

c) Land use in Houston

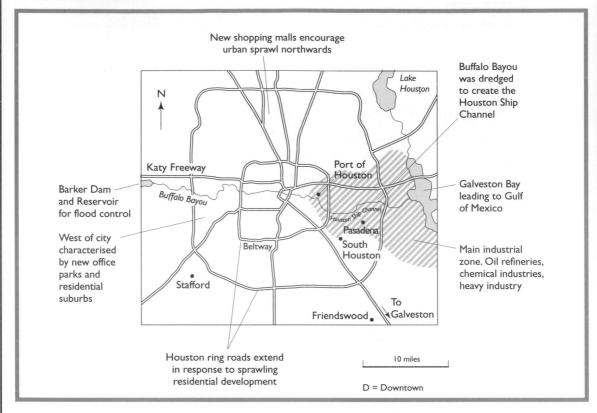

New shopping malls encourage urban sprawl northwards

Lake Houston

Buffalo Bayou was dredged to create the Houston Ship Channel

N

Katy Freeway

Port of Houston

Barker Dam and Reservoir for flood control

Buffalo Bayou

Galveston Bay leading to Gulf of Mexico

Houston Ship Channel

West of city characterised by new office parks and residential suburbs

Pasadena

South Houston

Beltway

Main industrial zone. Oil refineries, chemical industries, heavy industry

Stafford

To Galveston

Friendswood

Houston ring roads extend in response to sprawling residential development

10 miles

D = Downtown

d) Population

A growing economy attracts many migrants. Houston does not have the racial segregation typical of other American cities, although there is some concentration of whites in suburbs and immigrant groups downtown. The population is one-third white, one-third black and one-third brown – the mixture of blacks, Latinos, Asians and whites creates an enviable racial balance.

However, immigrant groups have a poor record in education in an environment where a good job depends on having a good degree. There is a high drop-out rate from schools: Latinos 50 per cent; blacks 20 per cent; whites 10 per cent. Whites are moving into private schools and abandoning public education, a process which will reduce integration and reinforce negative attitudes in state schools.

e) Houston – the unplanned city

There are no land-use planning restrictions in Houston and little red tape. Consequently, planning decisions are made by developers and local residents. Neighbourhoods are a mixture of residential and commercial land use. Small plots of land near the city centre are left in favour of vast planned low-density communities further out. This process keeps land values and density low – 2300 people/ha compared with 2000 in New York – but some suburbs are now 50km from Downtown.

f) Going up: the market is showing growth

New home in Texas

g) Office parks in Houston

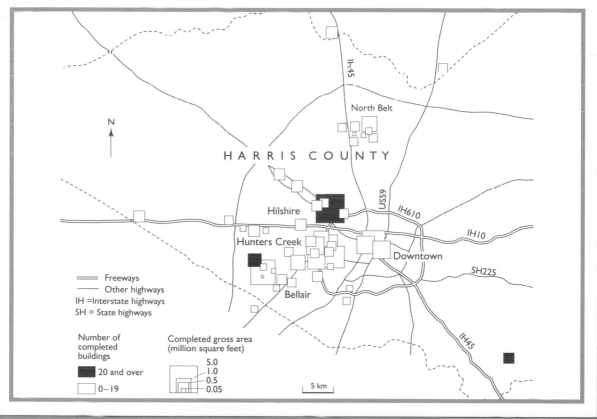

Legend:

Freeways
Other highways
IH = Interstate highways
SH = State highways

Number of completed buildings
■ 20 and over
□ 0–19

Completed gross area (million square feet)
5.0
1.0
0.5
0.05

5 km

Map labels: IH45, North Belt, HARRIS COUNTY, Hilshire, US59, IH610, IH10, Hunters Creek, Downtown, SH225, Bellair, IH45, N

ITEM 2
The economy of Houston

A city of many talents

- Second largest port in America, despite being 80km from the sea
- Home to Compaq, America's third largest PC maker
- NASA (the National Aeronautical Space Administration) is based in Houston
- Texas Medical centre, the largest medical complex in the world treating 70 000 patients a day is in Houston
- An ability to create new firms – Houston has created more new companies than any city in the US

Indicators of success in the year 2000

- Employment growth of 63 000 new jobs, with 81 000 forecast for 2001 – this makes Houston the second fastest city for job growth
- Per capita income up by 3 per cent
- Substantial increase in home ownership in urban and suburban communities
- Building permits issued for $4 billion in construction activity, an all-time record

Economic diversification – Houston has a broad spectrum of industries

- Oil and gas exploration
- Petroleum refining and production
- Medical research
- High technology industries
- Commercial fishing
- Agriculture
- Banking and finance

Energy: the driving force

Energy, and particularly oil, accounts for over half of Houston's economy but, even in the slump of 1999, Houston still managed to increase its number of jobs. Future prospects look bright as the US is forced to increase its energy production, including electricity, in the next 15 years. New technology in oil exploration and development requires high-level computing technology and huge investments, accessible only to the largest power companies. Houston is becoming the centre of US energy financing.

Enron, Houston's most successful energy company, has never explored for, pumped up or refined a single drop of oil, but was a key player in financing energy exploration before its bankruptcy.

ITEM 3

Environmental problems in Houston

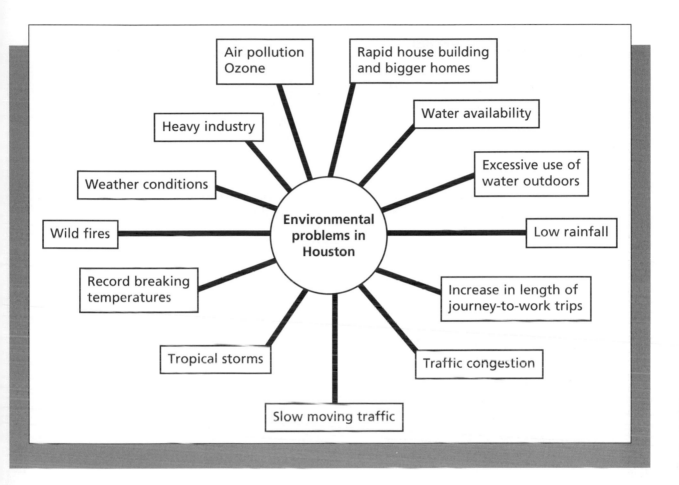

ITEM 4
Houston headlines

a) Plan to curb Houston's pollution

Houston has many sunny, calm days in winter and in summer, plus land and sea breezes from the Gulf of Mexico. These conditions, combined with extremely high traffic pollutants, lead to excessive levels of ozone and nitrogen oxides.

The state's environmental agency has approved an aggressive clean-air plan for the Houston metropolitan area. This smoggiest US city, with its refineries and petrochemical plants, surpasses even Los Angeles. The plan requires the Houston area industrial plants to reduce their smog-causing nitrogen-oxide releases by an average of 90%. However, large companies have argued that cuts in emissions of more than 75% are not technically or economically feasible.

Also under the plan, speed limits would be reduced to 55mph but taxi drivers are complaining that they will lose half their income by driving slowly. Other aspects of the plan include stricter exhaust tests for cars and trucks; a morning ban on the use of diesel construction equipment during part of the year; the sale of cleaner diesel fuel; and the replacement of old diesel equipment.

Some people think the proposals don't go far enough. Left out of the plan were proposals for 'no-drive days' and a requirement for home owners to overhaul ozone-eating air conditioning systems.

Associated Press, 7 December 2000

b) 'Water Smart' conservation campaign, June 2000

The campaign aims to raise awareness of problems caused by extreme events such as floods, tornadoes, storms and droughts. Population growth and rising expectations of the availability of water for swimming pools, gardens, etc. are leading to the risk of water shortages in south-east Texas. As temperatures rise and rainfall diminishes there are problems in keeping up with demand. Many reservoirs are only half full at the beginning of summer. Public infrastructure needs $13.1 billion investment to meet the requirements for safe drinking water in the region. The continued sprawl of Houston adds to the problems of water supply provision. Houston residents are urged to reduce outdoor water use in order to safeguard drinking water.

c) Wild fires cause smoke pollution in Houston, September 2000

Large widespread forest and marsh grass fires in south-east Texas resulted in heavy smoke pollution in the Houston area. Prolonged dry conditions and record-breaking temperatures significantly increased fire risks. A large haze developed from a mixture of ozone and smoke from industrial activities, urban areas and fires in rural areas. The young and elderly and those with respiratory or heart ailments were warned to take particular care.

d) Tropical Storm Allison

Houston is susceptible to tropical storms which blow in from the Gulf of Mexico and cause flash flooding in the metropolitan area. In June 2001, Tropical Storm Allison brought 300mm of continuous, excessive rain to neighbourhoods in south-east Houston. Massive street and roadway flooding resulted as creeks overflowed and waters on the low-lying land took time to recede. Electricity and telephone supplies were cut for two days, parking lots and basement offices were flooded and motorists were rescued from cars as their vehicles stalled in the rapidly rising waters.

There were 85 000 flood victims, 48 000 homes affected, 3500 homes destroyed and $497 million allocated to disaster assistance.

e) The Katy Freeway

The Katy Freeway extends 65km west from Houston's CBD to the Brazos River. Opened in 1968, it was designed to carry 79 200 vehicles per day and last for 20 years. Now, 30 years later, Katy Freeway carries 207 000 vehicles per day creating congestion for 11 hours per day, even at weekends, and is a deterrent to conducting business in the Houston area. Maintenance costs are $7.9 million per year, that's $121,500 per kilometre and four times the cost of other roads. The good news is that, at last, and with business and community support, there is a plan to improve the Katy Freeway – but the congestion while it is being built will be hard to bear.

a Read the Introduction and annotate the map (Item 1) to identify why Bintulu has a high population growth rate.   **8 marks**

b To what extent has the growth of Bintulu been dependent on:
  i)   physical resources?
  ii) human resources?   **18 marks**

c Outline the social, economic and environmental problems which may arise as a result of rapid development in the state of Sarawak.   **14 marks**

d What issues does the Sarawak government need to address in order to ensure sustainable development?   **10 marks**

Total 50 marks

Introduction

Sarawak is the largest state in Malaysia covering about 37 per cent of the total land area. Sarawak is one of the two Malaysian states on the island of Borneo.

Bintulu is a coastal town located in central Sarawak, and the administrative centre of Bintulu Division. It is about 650km from the state capital, Kuching, and about 200km from both Sibu and Miri, the other two main towns. Bintulu has the highest annual growth rate, 3.6 per cent, of all the divisions in Sarawak.

Bintulu can be reached by air from Kuching, which has direct connections to Singapore, Kuala Lumpur, Manila, Seoul, Taipei and Tokyo. Its $70 million all-weather deep-water port enables direct services with major ports in the region. The only all-weather route through the state, the Sarawak Highway, links Bintulu with the north and south of Sarawak.

The Bintulu coastal area is generally flat scrub grassland, with some mangrove on the coast. The terrain gets hilly towards the interior, where the natural vegetation is tropical rainforest although much has been cleared for oil palm plantations.

Economic development in Bintulu is focused around training a skilled workforce and developing a wide range of tourist facilities.

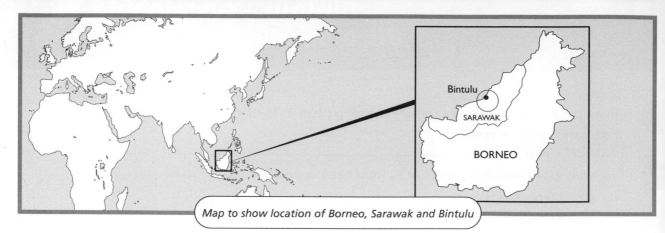

Map to show location of Borneo, Sarawak and Bintulu

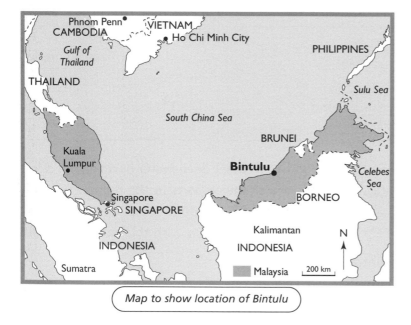

Map to show location of Bintulu

ITEM 1

Location of Bintulu

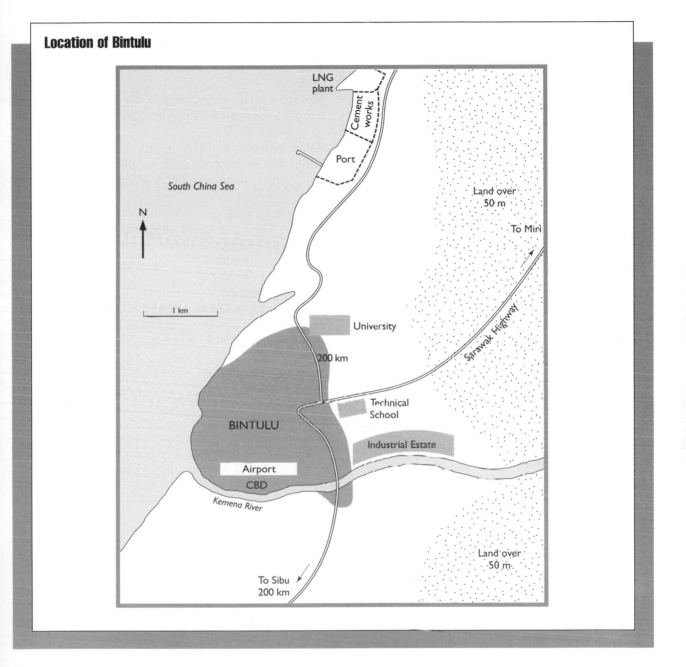

ITEM 2
Economic development of Bintulu

a)

The development of Bintulu as the leading industrial centre on Sarawak follows the discovery of large offshore natural gas reserves in 1968 and the building of Sarawak's first deep-water port. Although the immediate attractions are in the petrochemical and gas related industries, Bintulu is also an ideal site for a number of other resource-based industries as well as tourism. There is an abundance of coal, silica sand deposits, agricultural land for large-scale cultivation and potential for large-scale tropical timber development. Bintulu also has the capacity to support energy-intensive industries like aluminium smelters and steel plants, once the Bakun hydroelectric power plant starts to operate within 10 years.

The Bintulu Development Authority (BDA) is a development co-ordinating agency set up by the Sarawak government to help investors to establish industries and other business ventures in Bintulu. The Authority has provided infrastructure as well as sites for various industries.

b) Main industries in Bintulu

Malaysia Liquefied Natural Gas plant (MLNG)
Malaysia is currently the world's third largest exporter of liquefied natural gas (LNG).

Asian Bintulu Fertilizer plant
The plant was established in 1980 and is among the largest fertiliser plants in Asia.

Shell Gas plant
Built in 1993, the plant converts natural gas into high quality synthetic oil products and speciality chemicals.

Palm oil storage
This tank farm for local palm oil is connected to the general cargo terminal of the deepwater port by pipes.

Palm oil refinery
Produces edible oils.

Glue factory
Produces resin for plywood and chipboard manufacturing and timber processing industries, and supplies the Asian Bintulu Fertilizer plant.

Agricultural land in northern Sarawak		
Total area (hectares)	Percentage of total land area	Description
250 000	18%	Moderate slopes including flood plains
200 000	16%	Much wetland and areas of peat swamps – could be drained to improve land
820 000	66%	Steep slopes and severe erosion risk problems

Clinker grinding plant
The Bintulu cement plant produces both Portland Cement and blast furnace cement.

Medium Density Fibreboard (MDF) plant
The factory fulfils one of the State government objectives of utilising wood residues as raw materials.

c) Resources around Bintulu

Tropical timber
About 2.5 million hectares of Sarawak land is under forest, of which 30 per cent is in the Bintulu region. In the past, almost all the logs were exported with little processing done locally. However, the government has imposed a quota on the volume of logs to be exported, to encourage local timber processing activities. As a result, there has been an upsurge in investment in the timber processing industries in Bintulu.

Plantation and agro-based projects
Following the opening up of land along the Bintulu–Miri Road in the mid-1970s, the Sarawak Land Development started large-scale plantations for oil palm and cocoa within the Bintulu hinterland. The northern region of Sarawak including Bintulu is the most fertile and suitable for large-scale agricultural development: 500 000ha is for plantations. Currently 60 000ha have been cultivated with oil palm, 2000ha with rattan and 1000ha with pepper.

Natural gas
Sarawak has substantial gas reserves, 85 per cent of which is situated offshore near Bintulu. More gas discoveries are expected to be found. There is enough gas to meet current gas projects and the rest could be utilised for development of petrochemical industries.

Crude oil
Sarawak has a total known oil reserve of about 0.8 billion barrels of which 50 per cent are situated offshore from Bintulu. The reserve is mostly located in the producing fields situated about 30 to 40km off the Bintulu coast.

Hydroelectric power
For the time being, natural gas is used to run turbines to produce electricity. Sarawak has undeveloped hydroelectric power generating potential at 51 sites with a total combined capacity of 20 000 megawatts (MW).

Coal
Coking coal deposits of economic potential have been discovered in several areas of the Bintulu region. Each area has a probable reserve ranging from 750 000 tonnes to about 12 million tonnes.

d) Types of cargo leaving Bintulu port 1991–2000

Type of cargo	Number of ships, 1991	Number of ships, 2001
1. LNG	134	306
2. General cargo	538	914
3. Logs	256	41
4. Petroleum products	285	567
5. Crude oil	66	144
6. Palm oil	28	117
7. Silica sand	204	409
8. Ammonia	52	59
9. Fertilisers	76	102
10. Supply boats	316	274
11. Others	103	165
Total	2058	3098

ITEM 3
The 1990s property boom in Sarawak

Since the late 1980s, the demand for space for various human activities in Bintulu has increased substantially for many reasons.

- Rising population and declining household size (from 5.77 in 1980 to 4.95 in 1991 and estimated 4.58 in 1997)
- Growing affluence of the population resulting from sustained strong economic growth (Sarawak achieved 9.8 per cent growth in 1995)
- Changing demands for better living, working and shopping environments as well as for more leisure, recreational and resort destinations
- Increased car ownership and mobility
- Improvements of existing and completion of new basic infrastructure facilities and industrial sites
- Diversification of economic base, shifting from primary sector to secondary and tertiary sectors, with emphasis on the development of export-led industries
- Influx of foreign direct investments and national companies into Sarawak
- Increased corporate activities in Sarawak

ITEM 4
Environment around Bintulu

Beaches and scenery

The most attractive coastline around Bintulu is north of the town. It consists of small inlets, beautiful and unspoilt golden beaches, rock promontories and coral reefs. These small inlets and beaches provide excellent avenues for water sports with great tourist potential.

The scenic Kemena River passes through luxuriant tropical vegetation, tiny farming hamlets with traditional longhouses scattered along the river banks.

National parks

Two of Sarawak's seven national parks are found near Bintulu.

■ Similajau National Park

The park, 20km from Bintulu, consists of a narrow rocky shoreline and an interior of dense primary rainforest. Simple accommodation and a ranger service are provided.

■ Niah National Park

Parts of the park are honeycombed with limestone caves containing a rich variety of flora and fauna including thousands of swiftlets which produce edible bird nests (sold as a high value product to China and Japan). The west mouth of the Niah Caves is one of the most important archaeological sites in South East Asia.

Tropical vegetation around Kemena River, Sarawak

ITEM 5
Future development in Bintulu

The Sarawak government has identified the following as having potential for development in the area around Bintulu:

Petrochemical and gas related

Liquid Petroleum Gas, methanol, ammonia/urea, oil refining, compound fertiliser, nitric acid/ammonium nitrate.

Timber based

Integrated timber industry comprising the following: modern sawmill, laminated board, moulding plant, drying kilns, veneer/plywood, manufacture of prefabricated houses, chipboard, fibreboard, furniture, timber preservation plants, joinery factories, pulp and paper mills.

Plantation and agro-based

Large-scale cultivation of oil palm, and commercial farming.

Energy-intensive industries

Basic metal including iron, smelting and steel plants.

Other industries

Manufacturing of building materials such as specialised bricks and roofing tiles, food processing and shipyard industries.

Tourism

Tour operations – provision of tour packages for inbound tourists.

Provision of hotels, transportation, food, etc.

Tourist resorts – chalet, campsite and other facilities.

Boatels and establishment of sea sports centre for scuba diving, snorkelling, swimming, waterskiing, windsurfing, jetskiing and other specialised forms of sea sports.

Development of other tourism products such as traditional longhouses and Malay/Melanau cultural heritage and experiences.

Managing the environment – changing strategies in the Tongass National Forest

a Suggest why the Tongass National Forest region is considered to be of particular environmental value. **8 marks**

b Assess the impact on the physical landscape of
i) forestry
ii) mining **20 marks**

c Examine the economic and social implications of the decline of primary industries in Tongass. **14 marks**

d Suggest reasons why there is little opposition to the protection order on the Tongass National Forest. **8 marks**

Total 50 marks

Introduction

The Tongass is America's largest national forest (7 million hectares), located in the narrow east corner of Alaska, between the Gulf of Alaska and Canada. The dramatic landscape boasts an archipelago of glacial fjords, snow-capped mountains up to 3400m, tundra meadows and deep, steep-sided canyons. Mild, wet winters and cool humid summers are typical of the maritime climate where the high rainfall (3800mm) encourages the growth of a unique ecosystem. The most productive coniferous forests in the world, mainly hemlock and sitka spruce, line the lower slopes of glaciated valleys. Vast mineral resources also lie beneath the Tongass National Forest.

There has been US government approval of exploitation in the Tongass but its isolation has enabled a remarkable variety of mammals, birds and fish to flourish such as wolves, bears, eagles and salmon.

Since September 2001, the region has become the focus of controversy as the Alaskan Rainforest Conservation Act aimed to protect the region from timber and mineral exploitation.

ITEM 1
The Tongass environment

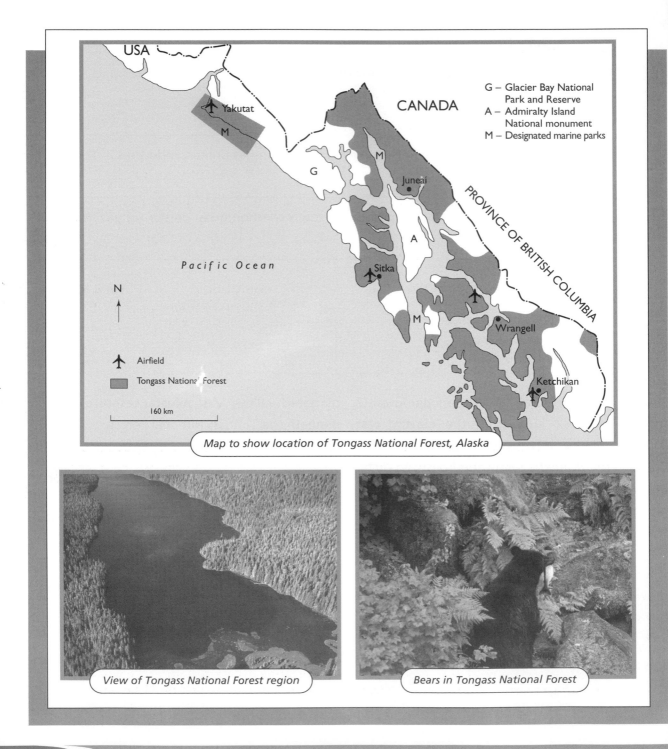

G – Glacier Bay National
    Park and Reserve
A – Admiralty Island
    National monument
M – Designated marine parks

Map to show location of Tongass National Forest, Alaska

View of Tongass National Forest region

Bears in Tongass National Forest

ITEM 2
The Tongass timber industry

a) Characteristics

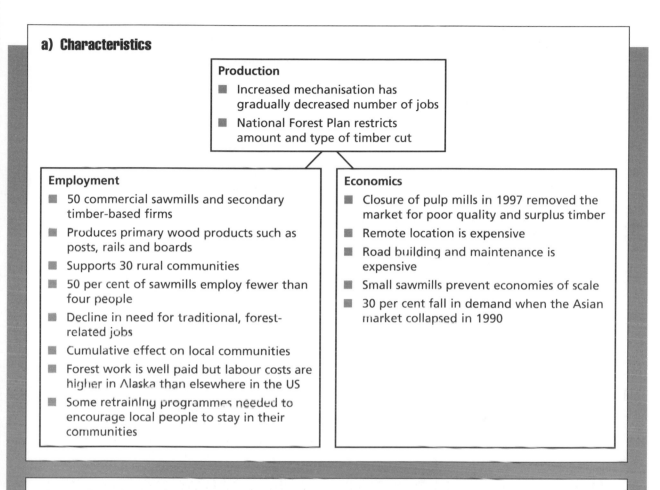

Production
- Increased mechanisation has gradually decreased number of jobs
- National Forest Plan restricts amount and type of timber cut

Employment
- 50 commercial sawmills and secondary timber-based firms
- Produces primary wood products such as posts, rails and boards
- Supports 30 rural communities
- 50 per cent of sawmills employ fewer than four people
- Decline in need for traditional, forest-related jobs
- Cumulative effect on local communities
- Forest work is well paid but labour costs are higher in Alaska than elsewhere in the US
- Some retraining programmes needed to encourage local people to stay in their communities

Economics
- Closure of pulp mills in 1997 removed the market for poor quality and surplus timber
- Remote location is expensive
- Road building and maintenance is expensive
- Small sawmills prevent economies of scale
- 30 per cent fall in demand when the Asian market collapsed in 1990

b) Benefits of the forest industry

- Clear-cut logging allows young trees the opportunity to thrive in maximised light conditions with minimum competition.
- Natural regeneration is rapid in Alaska's coastal forests.
- Different species require different habitats. Harvesting adds to this diversity. Often, wildlife increases and flourishes after harvesting. Where there is game, there are predators. Bear, wolf and human hunters, find excellent herds of deer, moose and other browsing species.
- Debris in streams was considered harmful to fish. Evidence now that it helps provide habitats where fish can hide and spawn.
- Each direct timber job creates at least three indirect jobs for truck drivers, road builders, doctors, retailers, teachers, and more.
- Each American consumes 300kg annually of wood products.
- Currently, forest growth in the United States exceeds harvest by 37 per cent.

ITEM 3
Mining in the Tongass

a) Mineral resources

Vast mineral resources lie beneath the Tongass National Forest.

- Greens Creek is the largest silver producer in the nation.
- Quartz Hill contains 11 per cent of the world's molybdenum resources.
- Kensington contains 50 tonnes of gold in 13 million tonnes of ore.
- A-J mine near Juneau has 120 tonnes of gold in 100 million tonnes of ore, and may hold four times that.

Forest Service planners argue that minerals must be included in plans for management. Their plan shows areas with the best prospects for development in the next 15 years. Foresters and miners have worked closely together to argue that miners could often use roads and sites that loggers had already cleared so not disturbing other places valued for wildlife and recreational use.

b) Environmental damage

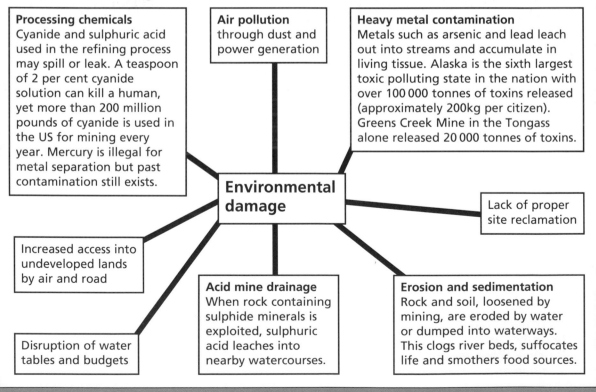

Processing chemicals
Cyanide and sulphuric acid used in the refining process may spill or leak. A teaspoon of 2 per cent cyanide solution can kill a human, yet more than 200 million pounds of cyanide is used in the US for mining every year. Mercury is illegal for metal separation but past contamination still exists.

Air pollution
through dust and power generation

Heavy metal contamination
Metals such as arsenic and lead leach out into streams and accumulate in living tissue. Alaska is the sixth largest toxic polluting state in the nation with over 100 000 tonnes of toxins released (approximately 200kg per citizen). Greens Creek Mine in the Tongass alone released 20 000 tonnes of toxins.

Environmental damage

Lack of proper site reclamation

Increased access into undeveloped lands by air and road

Disruption of water tables and budgets

Acid mine drainage
When rock containing sulphide minerals is exploited, sulphuric acid leaches into nearby watercourses.

Erosion and sedimentation
Rock and soil, loosened by mining, are eroded by water or dumped into waterways. This clogs river beds, suffocates life and smothers food sources.

c) Greens Creek mine, Juneau, Alaska

With a projected 17-year mine life, Greens Creek is one of the nation's largest silver producers. The operation has a mining and processing rate of 1300 tonnes per day. In 1997, the operation produced approximately 20 tonnes of silver, 1 tonne of gold, 34 000 tonnes of zinc and 15 000 tonnes of lead. These concentrates are shipped to various smelters throughout the world.

Current reserves are 200 tonnes of silver, 15 tonnes of gold, 675 000 tonnes of zinc and 225 000 tonnes of lead.

Tongass is home for the world's largest concentrations of bald eagles, grizzly bears, and Sitka black-tailed deer. Greens Creek Mine environmental programme works with local, state and federal government agencies to ensure that the wilderness of Tongass is preserved for future generations.

Juneau Harbour, Alaska

Greens Creek mine, Juneau, Alaska

ITEM 4
Local economies

a) Juneau

About Juneau	Percentage of population in industries	
■ Main administrative centre for Tongass	Forestry, fishing, farming	4.2
■ Summer tourism important: $130 million and 2000 jobs	Mining	3.0
■ $80 million from cruise ships docking – trips to ice fields and glaciers	Construction, transport, communications	13.7
■ Support services to logging and mining industries	Manufacturing	2.5
■ Kensington Gold is being developed; Greens Creek also significant		
■ Fishing earned $10.4 million – salmon hatchery and cold storage facilities for fish processing	Retail and finance	25.4
	Recreation, public services, administration	51.3

b) Ketchikan

About Ketchikan	Percentage of population in industries	
■ Industrial centre and major port	Forestry, fishing, farming	5.8
■ 500 000 visitors from cruise ships	Mining	0
■ Large fishing fleet, fish processing facilities, timber and wood processing	Construction, transport, communications	18.3
■ Ketchikan Pulp Corporation pulp mill closed 1997, 400 jobs lost	Manufacturing	13.2
■ Several small companies developing – specialist wood veneers		
■ $10 million earned from fishing – fish hatcheries, cold storage and fish processing	Retail and finance	24.7
	Recreation, public services, administration	38.0

c) Wrangell

About Wrangell	Percentage of population in industries	
■ Economy based on commercial fishing and timber from Tongass	Forestry, fishing, farming	13.3
■ Some independent tourists – sports fishing on Stikine River	Mining	0.6
■ Alaska Pulp Corporation sawmill closed 1994 – 225 jobs lost, 20 per cent of workforce	Construction, transport, communications	16.7
■ Silver Bay Logging reopened mill in 1998 – 33 jobs		
■ Transport and support services for renewed gold mining on Stikine River	Manufacturing	16.7
■ $5 million revenue from fishing	Retail and finance	28.3
	Recreation, public services, administration	24.4

ITEM 5
The future for Tongass

a)

Until its closure in 1997, the Ketchikan Pulp Company's 50-year contracts forced unsustainable management on the Tongass. On Prince of Wales Island, which has few un-logged areas left, the long-term viability of wildlife populations, and tourism and recreational uses, are limited by the destruction of wilderness. Following the closure of the pulp mill, the Tongass Conservation Society will work towards sustainable harvesting of timber. A flourishing diverse timber industry could be developed, based on small businesses providing high-quality finished wood products that provide more skilled jobs using less timber. This could be combined with tourism, which would encourage diversification in local communities.

b) Tourism in Tongass

CRUISE SHIP TOURISM INCREASED IN 2001

Ketchikan saw a rise in the number of cruise ship passengers visiting its port this summer, but that trend, along with a growth in tourism-generated revenue, could be threatened by expansion of Canadian ports along the coast.

More than 665 000 of the 690 000 cruise ship passengers who visited Alaska stopped in the First City, according to the Ketchikan Visitors Bureau. That's an increase of more than 92 000 visitors from 2000.

Ketchikan Daily News, October 2001

c)

CLINTON GIFT OF ALASKAN WILDERNESS OUTRAGES US LOGGERS

In the final days of President Clinton's administration, he issued an order banning road construction, logging and mining on 30 per cent of US national forest. The 9 million acres affected in Tongass will deny logging companies access to the most profitable remaining stocks of spruce, cedar and hemlock trees. The region contrasts starkly with neighbouring British Columbia where all traces of rainforest have been eradicated by more than a century of exploitation.

But this is not entirely a straightforward fight between conservationists and capitalists. The logging that has been permitted in the Tongass since the 1950s has not really been about big corporations making money; rather it has been about providing economic resources and jobs, with heavy government subsidies, to inhabitants of remote rural areas.

If companies are keen to continue logging, it is largely because there are about $30 million of annual federal grants up for grabs. The operation makes little or no economic sense.

Two major pulp mills in Ketchikan and Sitka have closed down, turning the local logging industry into little more than a government welfare programme – and one that nevertheless threatens an irreplaceable natural resource.

"Many of the inhabitants understand that logging has no future – particularly since it threatens almost every other economic activity in the area from salmon fishing to tourism." Alaskan politicians see things differently – because the loggers and their companies are voters and campaign contributors.

The Independent, January 2001

a Item 1 describes the main features of the proposed Arsenal stadium development. Annotate a copy of Item 1c to identify the land-use changes which will result from the stadium redevelopment. **8 marks**

b Examine how the new development might locally benefit
i) the economic environment
ii) the social environment **16 marks**

c Describe the transport issues which need to be addressed in Islington if the stadium development goes ahead. **8 marks**

d Suggest reasons why some local residents and Arsenal fans oppose the new stadium development. **10 marks**

e Should the new Arsenal stadium be built? Justify your view. **8 marks**

Total 50 marks

Introduction

Arsenal Football Club has played at Highbury stadium in Islington, north London, since 1913. One of the oldest and now most successful clubs in the country, Arsenal needs to expand the stadium to cater for more fans and improve facilities. The current pitch is the smallest in the Premier League and the stadium holds only 38 500 people – no longer sufficient to cater for demand; Arsenal matches played at Wembley have sold 70 000 tickets. Although comprehensive redevelopment plans were investigated, the current site is too small to meet the needs of the club at national and international levels.

Arsenal wishes to build a new £250 million, 60 000 seater stadium in Ashburton Grove, 1km from Highbury stadium on the site of small businesses and a waste disposal facility.

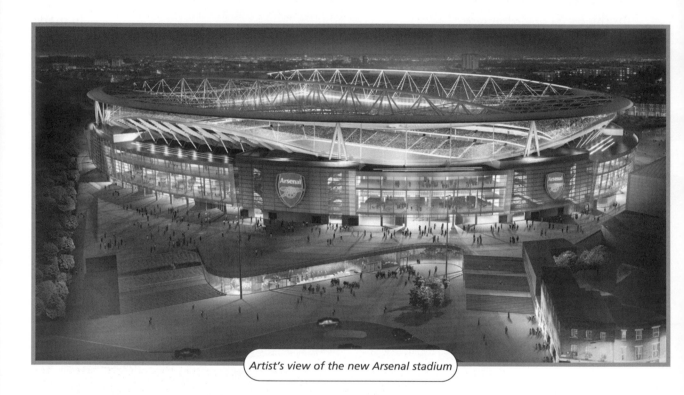

Artist's view of the new Arsenal stadium

ITEM 1

The plans for the new Arsenal stadium

a) The site of the new development

Highbury stadium is an important local landmark, surrounded by flats and terraced housing. The East Stand, a listed building, and the West Stand will be converted into flats with the North and South stands replaced by residential blocks; 25 per cent of new housing will be 'affordable' for local people. The pitch itself will be preserved as a landscaped open space for the private use of residents.

The existing sports and community centre will be demolished and replaced with two terraces of housing.

Ashburton Grove is well served by public transport and within walking distance of six Underground or rail stations. It is currently occupied by a range of service and industrial uses, and a borough cleansing and highways depot station, all of which generate substantial volumes of heavy traffic. These units would have to be relocated to a new site in Lough Road. Nearby there is commercial land use along Holloway Road plus the University of North London campus and residential areas. The site is affected by the existing Highbury stadium on match days, mainly from parking.

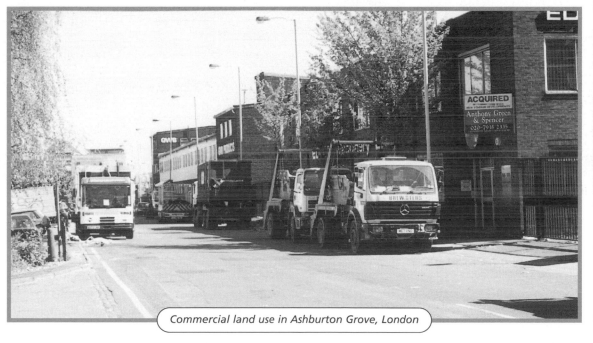

Commercial land use in Ashburton Grove, London

Lough Road is intended for the relocation of the waste disposal depot and other industrial uses currently at Ashburton Grove. Lough Road is a visibly run-down, contaminated industrial site with vacant buildings, builder's merchants, a mattress factory and brush factory, as well as a school and row of railwayman's cottages. Despite providing some employment it detracts from the socio-economic well-being of the area, and the London Borough of Islington has been seeking its redevelopment for some time. Plans include new, purpose-built offices and industrial units with a new access road, a refurbished listed building, new housing and student accommodation.

b) The stadium

The **new stadium** will comprise a 'podium' area 6m above street level. Beneath the podium area will be the club shop, museum and sports bar. There will be a six-storey building for club offices, multi-purpose sports hall for the sports and community centre, and commercial ground-floor space for pubs and restaurants, and underground parking.

Artist's impression of the new stadium

At **Highbury stadium**, Arsenal currently provides 150 full-time and 300 casual jobs rising to 800 on match days. It is one of the largest employers in Islington. The demand for casual staff is particularly important because this tends to benefit the less economically active population.

The main routes to Highbury are from Underground and main line stations – Finsbury Park, Drayton Park, Highbury and Islington.

There is temporary crowding of the area before matches which can be intimidating and unpleasant for residents, and there is some littering and anti-social behaviour. However, some residents set up stalls in their front gardens selling food, drinks and souvenirs on match days.

c) Map to show location of the three areas involved in the Arsenal project

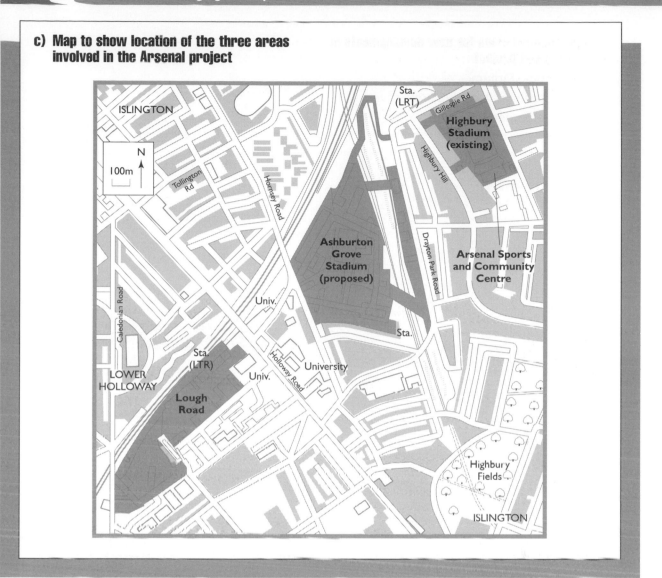

d) Map to show plans for new developments in the Arsenal project

i)   Highbury stadium

ii) New stadium at Ashburton Grove

iii) Redevelopment of Lough Road

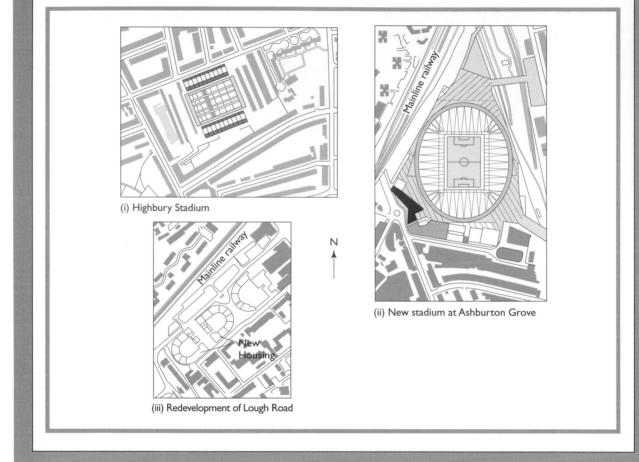

(i) Highbury Stadium

(iii) Redevelopment of Lough Road

(ii) New stadium at Ashburton Grove

ITEM 2
Islington facts and figures

a) Population

- Eighth most deprived local authority in England
- Wards of Highbury and Holloway are in the national top 20 per cent for multiple deprivation, health and education
- 50 per cent of households have no car
- 70 per cent of households live in rented accommodation
- 9000 existing households are living in unsuitable housing in need of repair
- Unemployment rate is 20 per cent for men and 14 per cent for women (i.e. much higher than the national average)

Current and projected population of the development area		
	Number (1991)	Projected number by 2011, excluding development proposals
Infants and children (0–15 years)	45 600	52 000
Working age (16–64/59)	142 500	164 500
Retirement age	29 500	31 500
1996 estimated population	**217 600**	
2011 estimated population		**248 000**

b) Employment

Long-term impact on employment in Islington if the development goes ahead	
Employment groups	% change
Manufacturing	−14%
Retail	9%
Hotel/catering	25%
Professional/business services	15%
Education/health	3%
Public	−8%
Other	21%
Total	14%

ITEM 3
Arsenal on the move

a) Urban regeneration

After extensive consultation, Arsenal Football Club believes that its proposals will act as a catalyst for the wider regeneration of the area by:

- creating 1800 new jobs
- retaining 2100 existing jobs
- providing 2300 new flats and houses with up to 35 per cent 'affordable' housing
- new residents bringing increased spending power
- relocation and replacement of old industrial/waste management premises with new buildings built to modern standards.

Open space

New areas of green and open space will be created by:

- extending the existing Gillespie Park ecological centre
- creating green space within the Highbury stadium redevelopment
- making an attractive open space around the new stadium.

b) Employment

Potential job creation at the new stadium		
Match day staff	Number of workers in existing stadium	Number of workers in new stadium
Stewards	350	900
Turnstile operators	80	20
Maintenance	11	16
Security	7	11
Cleaning	27	50
Ancillary other	45	100
Catering	300	1347
Total	**820**	**2444**
Non-match day staff		
Catering		215
Contracted personnel	151	186
Total	**151**	**401**

The Arsenal project could bring benefits to local retailers.

- The number of local residents will increase, therefore giving an estimated £18.3 million extra trade annually.
- The 1.5 million visitors to Highbury each year will generate trade for shops, bars and restaurants.
- People commuting to new jobs in the Highbury area will use local shops while travelling to and from work.

Islington Borough Council aims to establish sustainable economic growth (i.e. encourage development in growth sectors such as business services, ICT communications, leisure and hospitality rather than warehouses and the heavy manufacturing sector). Leisure is particularly important because it develops an evening economy, too. The business sector provides both high- and low-skilled job opportunities and, therefore, a good long-term balance of employment within the borough.

c) Community benefits

Local residents, young people, health and education services will benefit from a range of community initiatives:

- four new health facilities, 79 jobs and better local health care
- children's nursery at Highbury Stadium
- children's nursery at Lough Road
- learning centre in the new stadium
- sports and community centre.

ITEM 4
Transport issues

a)

Mode of transport	Current percentage of people using each type of transport	Predicted percentage of people using each type of transport
Underground	48	55.8
National rail	11	14
Car driver	13	5
Car passenger	18	7
Walk	4	6.8
Bus	3	4.9
Coach	2	3.7
Taxi	1	1.8
Other	0	1
Total	**100**	**100**

- 65 per cent of spectators use local pubs, cafés, restaurants and shops before attending a match.

- Arrival times at a match are staggered – 65 per cent of spectators are in the stadium area one hour before kick-off.

- Departure times are extremely concentrated. This puts the most pressure on transport services.

- 50 per cent of spectators would consider remaining longer in the stadium if better post-match entertainment were provided.

- A new controlled parking zone would result in 2000 fewer cars being driven to the local area on matchdays (4800 people in total).

b)

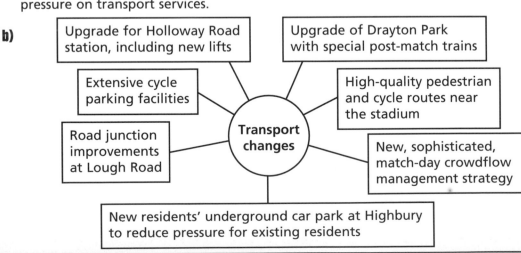

Upgrade for Holloway Road station, including new lifts

Upgrade of Drayton Park with special post-match trains

Extensive cycle parking facilities

High-quality pedestrian and cycle routes near the stadium

Road junction improvements at Lough Road

Transport changes

New, sophisticated, match-day crowdflow management strategy

New residents' underground car park at Highbury to reduce pressure for existing residents

c)

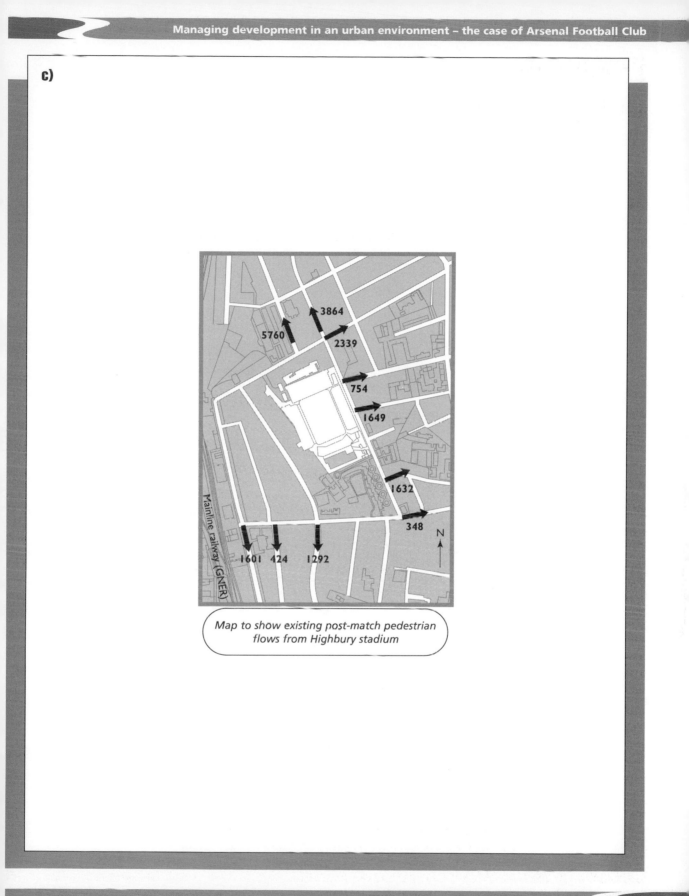

Map to show existing post-match pedestrian flows from Highbury stadium

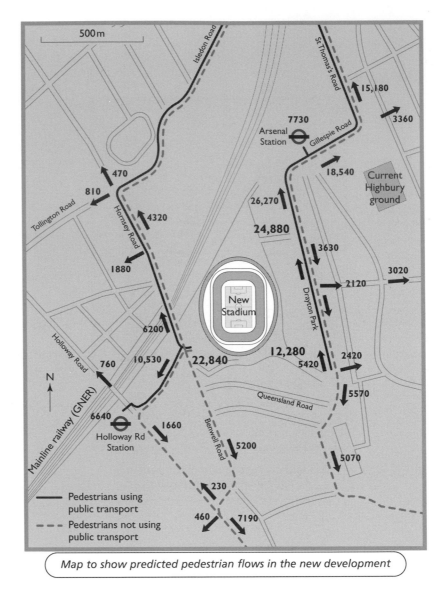

500m

Isledon Road

St Thomas's Road

15,180

3360

7730

Arsenal Station

Gillespie Road

18,540

Current Highbury ground

470

810

Tollington Road

Hornsey Road

4320

26,270

24,880

3630

1880

2120

3020

New Stadium

Drayton Park

6200

12,280

2420

Holloway Road

760

10,530

22,840

5420

5570

Mainline railway (GNER)

6640

Holloway Rd Station

1660

Benwell Road

Queensland Road

5200

5070

N

230

460

7190

— Pedestrians using public transport

- - - Pedestrians not using public transport

Map to show predicted pedestrian flows in the new development

ITEM 5
Local views about the Arsenal development

a)

Arsenal fans who have spent years cheering on their heroes from the East Stand could get the chance to move in for life. Property developers think it would make the perfect home for a devoted Gunners fan. 'People would be queuing up to move into the stadium. Highbury is an up-and-coming area even with the current stadium. When it's gone it will be even more desirable.'

The East Stand is the jewel in the crown of Highbury with the impressive marble hall where the trophy cabinet is kept. The development would also be boosted by plans to turn the pitch into a picturesque park so those who move in could relax on the hallowed turf. Gunners fan Nick Hallam said 'I would love to live at Highbury although I would need to win the lottery to afford it'.

Islington Gazette, May 1999

b) Residents oppose Arsenal stadium plan

Arsenal's plan to convert a rubbish dump into a new £100 million stadium is facing fierce opposition from local residents and businesses.

Although a majority of the firms located on the 25-acre plot of land are owned by the council, a number of privately run companies are reluctant to move. Arsenal fears they may try to hold the club to ransom, thereby forcing up the cost of the project.

Councillor Maureen Leigh said 'I have had some very negative responses. We all know what goes on around football grounds and the new plan would be a blot on the area. The primary concern is the effect on traffic congestion especially in Holloway Road. It's already under pressure when Arsenal play at Highbury.'

Arsenal will encourage fans to travel to the new stadium by tube by linking the sale of tube and match tickets.

c)

'The scheme looks likely to cost not increase jobs. Local people will lose jobs on site. Businesses will close. New premises will not have the guaranteed low rents they have at present.'

With an additional 22 000 fans making their way to matches, and limited plans to improve transport links, Nick believes the proposal runs counter to the local council's promises to reduce traffic. Other than re-siting the JVC Sports centre, there are few benefits to the local community. Simply keeping Arsenal in Islington doesn't count because many of fans come from outside the borough.

Understandably, rumours of a 'Chelsea Village' style development with hotel, casino, shopping mall, etc. could make things worse. Nick listed some of the issues worrying local residents – noise and light pollution, reduced access for emergency services, litter, street trading, disrupted TV reception, increased anti-social behaviour, unpleasant views and two years of building works.

Nick asks 'Why has no other club contemplated a stadium on this scale in a residential area just 60m from people's homes?'

Arseweb exclusive interview with Nick Robinson, BBC Radio 5 Live presenter

d) Other objections to the Arsenal plan

- Nose, light and air pollution
- Unacceptable litter
- Increased parking difficulties
- More traffic
- Under-estimation of traffic congestion in post-match hour
- Overloaded public transport
- Inadequate estimates of train availability
- Inadequate provision for large crowds at rail and Underground stations
- Loss of permanent jobs in local businesses
- Growth of fast-food outlets displacing local retailers
- Lack of public access to open space especially in Highbury stadium
- Insufficient affordable housing
- Huge structure of stadium inappropriate for residential area
- Difficulties for local pedestrians on match days

Itaipu – the costs and benefits of renewable energy

a

i) Using the data in Item 1a, draw a graph to show changes in the main energy sources in Brazil, 1979–2000   **8 marks**

ii) Comment on changes in the relative importance of different primary energy sources in Brazil between 1979 and 2000.   **8 marks**

b Compare the national benefits of the Itaipu HEP scheme for both Brazil and Paraguay.   **8 marks**

c Assess the **local** impacts of the Itaipu dam on the
i)   physical environment
ii) social, economic and political environment   **18 marks**

d Discuss why it is unlikely that dams the size of Itaipu will be built in the future.   **8 marks**

Total 50 marks

Introduction

Itaipu is the world's largest hydroelectric power plant, situated on the border of Paraguay and Brazil, and affecting 200km of the Parana River. The man-made reservoir, Itaipu Lake, is 8km long. It took 18 years to build at a cost of $US 18 billion and required the integration and co-operation of local, state and national governments plus local and multinational corporations. It is administered by Itaipu Binational, a joint Paraguayan and Brazilian government venture.

Each country is entitled to half of the energy produced from the dam. Brazil has the first option to buy power from Paraguay which uses only 3 per cent of Itaipu's energy.

Electricity was desperately needed to develop the industries of south-eastern Brazil, without which economic development would be limited. Critics argued that Brazil neither needed nor could afford such a massive hydroelectric scheme and that the country would have been much better served by smaller and less prestigious schemes nearer to centres of consumption.

Map to show location of Itaipu in South America

Map to show location of Itaipu on Parana River

ITEM 1
Sources of energy

a) Brazil

Brazil has a range of energy sources and needs cheap power to refine its considerable mineral resources. Rapid urban development in the south also demands power. Oil reserves are insufficient to meet Brazil's needs and ethanol from sugar cane was used as a petrol substitute.

Brazil, with energy-hungry industries and a population of over 130 million, has seen an increase in electricity consumption of 10 per cent in 30 years.

Primary sources of energy (% of total)					
Source	1979	1983	1988	1994	2000
Hydropower	25.1	30.4	33.8	37.8	46.0
Wood	22.3	20.7	17.3	12.7	7.7
Sugar cane	5.9	9.8	10.2	9.8	6.3
Oil	46.0	32.4	29.7	30.6	31.2
Natural gas	41.3	1.3	2.1	2.4	3.6
Coal	0.7	4.7	5.7	5.3	4.6
Nuclear	0.0	0.0	0.1	0.1	0.6

b) Paraguay

Paraguay has no coal, oil or natural gas and is therefore heavily dependent on HEP. It has more than enough energy for its needs and electricity accounts for 15 per cent of exports. The remaining export products are cotton, soy beans, timber, vegetable oils, meat products and coffee.

The joint Paraguay–Brazil commission overseeing Itaipu has accumulated $4.2 billion in debt. This is mainly as a result of electricity rate reductions decreed unilaterally by Brazil, the monopoly buyer, since the project was completed in 1991. The sale price of electricity has been below the production cost, subsidising Brazilian power companies which purchase 97 per cent of total output from Itaipu. Sales of energy to Brazil pay for 30 per cent of Paraguay's national budget and the country is still owed money from energy sales in the 1990s.

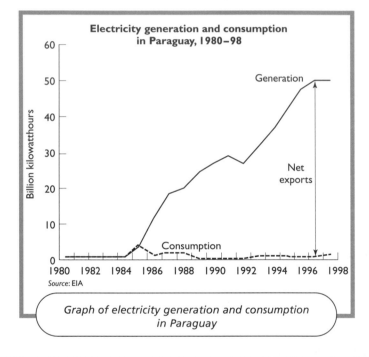

Graph of electricity generation and consumption in Paraguay

ITEM 2
Main features of the Itaipu dam

a)

River basin	
Area of Parana river basin	820 000km^3
Average annual precipitation	1400mm
Average discharge at Itaipu	9700 m^3/s
Reservoir	
Area at maximum normal level	1350km^2
Volume at maximum normal level	29 billion m^3
Length	170km
Average depth	22m
Islands	66 small islands created
Dam	
Height	196m
Length	7.5km
Turbines	18 (+ 2 more operational from 2004)
Generating capacity	12 600MW

b) The local environment

The Parana region has rich, fertile soils with natural semi-tropical rainforest. Intensive agriculture produces large amounts of grain, particularly soy beans. The submerged Parana River formerly flowed over several waterfalls, the Seven Falls described as one of nature's wonders, and then through a small canyon. The river used to be navigable for small vessels but the dam makes navigation impossible.

ITEM 3
Itaipu conservation programmes

The impact on the natural ecosystem around Itaipu was enormous, especially in the initial stages of dam construction. However, many endangered species were salvaged as Brazil and Paraguay established protection programmes, but others unique to Itaipu were destroyed.

Land has been classified as:

- prohibited area (1800ha) with no access
- environmental recuperation (19 700ha) no access except to the lake
- restricted use (6800ha) for ecological parks, nature trails, education
- intensive use (900ha) for tourism, leisure and economic development.

a) Gralha Azul project
Although forest protection was initially included in the planning proposals, 700km² of forest have been damaged, especially in Paraguay where 85 per cent of forest was destroyed. A 200m-wide forest belt borders Itaipu lake to help prevent silting up, erosion and pollution; 11.5 million plant species were transplanted and 105 000ha of land have now been protected.

The project aims to:

- protect rare species
- reforest areas affected by early dam construction
- create three 100ha forest reserves
- plan ecological and recreation development along river banks.

b) Mymba Kuera project
This project aimed to minimise the effects on fauna of reservoir flooding by catching animals and relocating them in biosphere reserves. As a result of this project, 27 150 animals, birds and insects were rescued.

Technicians rescuing animals before filling of Itaipu reservoir, Brazil

c) Soil conservation

In the agricultural areas around Itaipu lake, the construction of small hills (*murundus*) and contour curves aimed to stop torrential rains from taking tonnes of soil and chemicals into rivers which flow into the lake. This would help farmers maintain the quality of the soil.

Murundus *(small hills to stop erosion by rainfall) in Brazil*

d) Pesticide pollution

Itaipu National, local governments and co-operatives are recycling the 10 million wrappings of the pesticides used in agriculture. Dedicated ponds are available to wash equipment used in pesticide application. This will prevent farmers washing their equipment in lake waters.

e) Aquatic ecosystem

Above the dam there are more fish but fewer species, while below Itaipu there is more variety but less volume of fish. A 6km-long fish canal is being built to aid fish migration. Commercial fishing is productive (1600 tonnes per year) and artificial breeding maintains fish stocks.

Aquatic plants are liable to become weeds and affect leisure, navigation, fishing and irrigation when they proliferate. Transpiration increases water loss and dredging is required to remove the weeds.

Water quality is safe for swimming; the amount of heavy metals, pesticides and nutrients are within legal limits.

ITEM 4
Impacts of Itaipu

a) Social and economic issues

- Limited compensation payments for displaced communities
- Local transport along the river
- Increase of tourism and recreation
- Resettlement of Guarani people whose land was flooded on to reservations
- Disruption to lifestyles of rural dwellers
- Disruption to host population living in areas designated as reservations
- Inundation of archaeological sites
- Water-borne diseases
- Merging of rural and urban cultures
- Local people claim that the climate around the lake is more humid with more fog, and is leading to chest infections and ill health in local towns and villages

b) Energy crisis in Brazil

Despite the massive scale of Itaipu and other dams like it, Brazil has an energy shortage. The government blames the worst drought in 30 years, which led Brazil to become precariously dependent on HEP.

Itaipu is producing electricity at maximum capacity. However, there are insufficient power lines to transmit electricity from the southern state of Parana.

c) Brazilian fishermen suing operators of Itaipu, world's largest dam

by Glenn Switkes

In a David and Goliath battle, four fishermen's associations are suing the company that operates the world's largest dam, Itaipu, on the Paraná River on the border of Brazil and Paraguay. The associations took their case against Itaipu Binational, which manages the huge 126000-megawatt-capacity dam, to a federal district appeals court earlier this year. The groups say that the company's deliberate lowering of Itaipu's reservoir level to maximise power generation during dry periods has caused massive fish kills, adversely affecting their livelihood. Some 1000 artisanal fishermen earn a living from fishing in the Itaipu reservoir.

With Itaipu supplying 25 per cent of Brazil's electric power, the company opened the floodgates of the dam to maximise power generation during the dry season this year, resulting in a drastic drop in the level of the reservoir – up to five metres. This has caused the margins of the reservoir to recede by three kilometres in some places along its 160km extension. Coming as it does during the period of fish reproduction, the impact on fish populations has been significant.

High-value migratory fish, such as the dourado, pintado and cachara, have long-since vanished from the reservoir area, and now lower-value fish, such as the bottom-feeding armal, are affected by the death of millions of tiny crustaceans that normally are found on the reservoir's edge.

Itaipu was built before environmental laws were enacted in Brazil. In the 18 years since its inauguration, the dam has not been subject to environmental impact studies, nor required to obtain an environmental licence. Some 60000 people were relocated to make way for its construction. According to Aparecido da Silva Martins, lawyer for the fishermen, 'Itaipu is a relic of the military dictatorships, which chose most of its current directors. It is run as if it were a separate country – a country without laws.'

The fishermen's appeal follows a decision by a lower court judge that they cannot sue for personal losses because the river is a 'commons' which is available to the public for use, and is not their private property. In his opinion, the judge said that the dam has had significant impacts on fish stocks. The fishermen argue that they fish under federal licences, in the same manner that Itaipu operates as a federal concession. They say that this means that the company has no right to destroy their means of earning a livelihood, just as they would not be permitted to do anything to interrupt the functioning of the dam.

Itaipu Binational has responded to the claim by saying it is now constructing a canal which will permit migratory fish to bypass the dam. The company denies charges of a significant fish kill, and says it uses backhoes to dig temporary channels for fish that are trapped in dry areas, and that it even manually rescues some fish with nets. The fishermen respond that these efforts are symbolic, and have little effect on the broader impacts of Itaipu's reservoir management.

World Bank Rivers Review, October 2000

d) Power cut plunges Brazil into chaos

Alex Bellos in Rio de Janeiro

Brazil ground to a halt yesterday afternoon when up to 100m people were left without electricity in one of the largest power cuts in memory.

An unidentified problem at the Itaipu dam, on the border with Paraguay, cut off electricity for several hours in the populous south and south-east of the country.

The blackout caused chaos in Rio de Janeiro and Sao Paulo, Brazil's two largest cities, where traffic lights stopped working, both underground systems closed down and much commerce was forced to shut. Police rushed to take control of the situation.

The power cut happened at 1.30pm trapping thousands of people in lifts and plunging offices and factories into darkness. Power began to be reconnected after half-an-hour, but after two-and-a-half hours, 40% of Rio and Sao Paulo states were still lacking electricity.

Shortly after the blackout there was a 30-mile traffic jam in Sao Paolo. Mobile phone systems were unable to cope with the demand, since electrically powered conventional phones were not working.

Other large states, including Minas Gerais, Goias, Parana and Rio Grande do Sul were also affected – an area of about 500000 square miles.

Preliminary investigations into the cause of the incident pointed to a power line coming down at the Itaipu dam.

The president of the dam, Euclides Scalco, said the problem meant that all the complex's 18 turbines had to be switched off.

The power cut comes while Brazil is trying to cope with an energy rationing scheme to combat power shortages and avoid California-style blackouts. In recent weeks, however, higher than expected rainfall has raised water levels at the reservoirs which power Brazil's hydroelectric energy grid.

Guardian, 22 January 2002

a   i)   Describe and suggest reasons for the distribution of rainfall and temperature in Spain.

8 marks

   ii)   How has climate influenced the distribution of agriculture in Spain?   **6 marks**

b   Suggest why Spain has invested in so many dams along its rivers.   **6 marks**

c   Assess the potential hydrological and ecological impacts of the Ebro transfer scheme on the Ebro delta.   **10 marks**

d   Summarise the potential impacts of the Ebro transfer in the Alicante region.   **12 marks**

e   Does Spain need the Ebro Water Transfer Plan? Justify your view.   **8 marks**

Total 50 marks

Introduction

The climate of the Iberian Peninsula is subject to major year-on-year fluctuations, in particular drought and variable rainfall. There is a geographical disparity between the availability of, and demand for, water. Although Spain receives a net 110 billion m^3 of precipitation per year and uses only 35 billion m^3, the distribution of the water poses a fundamental problem for the government.

The irrigated areas of the Mediterranean coast and the inland river valleys are the most productive agricultural areas of Spain Sheltered, but receiving little rainfall, they produce fruit and vegetables when irrigated. The high levels of sunshine are also a valuable input and provide for the production of both early and exotic vegetables which are sold throughout Spain and the North European market.

The controversial Spanish National Hydrological Plan, supported by the government, proposes a major water diversion scheme from the Ebro River across several river catchments to Almeria. This plan aims to provide water to support the development of agriculture in the southern region and reduce regional disparities within Spain.

Map to show the regions of Spain

ITEM 1
Climate and agriculture in Spain

a) Relief map of Spain

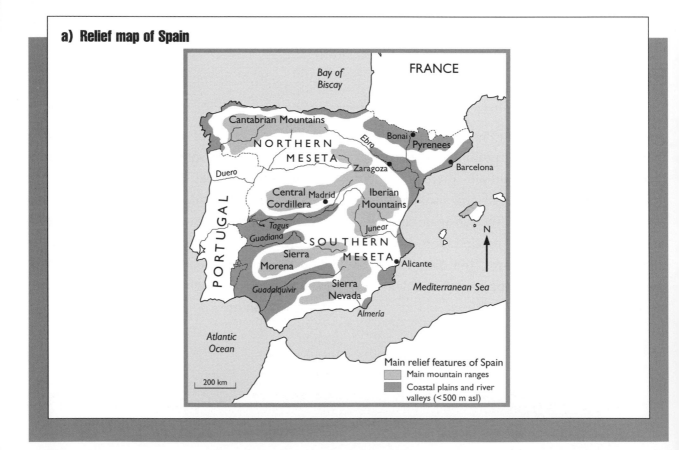

b) Rainfall in Spain

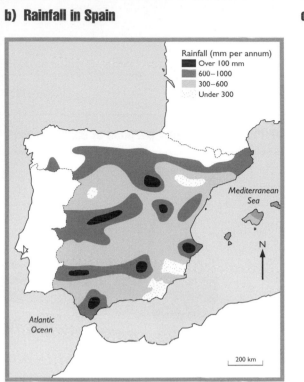

Rainfall (mm per annum)
- Over 100 mm
- 600–1000
- 300–600
- Under 300

Mediterranean Sea

Atlantic Ocean

200 km

c) July isotherms in Spain

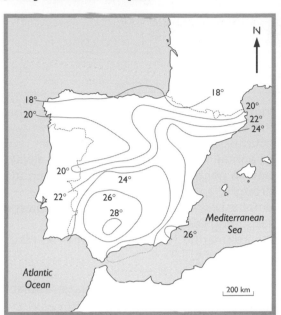

Mediterranean Sea

Atlantic Ocean

200 km

d) Distribution of agriculture in Spain

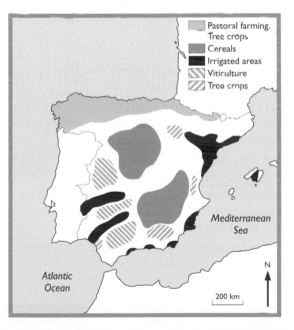

- Pastoral farming. Tree crops
- Cereals
- Irrigated areas
- Viticulture
- Tree crops

Mediterranean Sea

Atlantic Ocean

200 km

ITEM 2
Dams in Spain

In its attempts to conserve water, Spain has more dams than any other European country. Spanish dams regulate 40 per cent of annual surface runoff. Water has always been a political priority because of periodic drought, and government policy has emphasised large public works regardless of recouping costs. Economic and environmental factors have not been taken into account.

Spain now has 2500 litres of water available per person per day. This compares with 1800 litres in the US and 1200 litres in Israel.

a) Pie chart to show purpose of dams in Spain

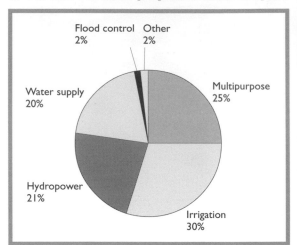

Flood control 2%
Other 2%
Multipurpose 25%
Water supply 20%
Hydropower 21%
Irrigation 30%

b) Use of water in Spain

Irrigation	Accounts for 80 per cent water consumption. Spain irrigates 3.4 million ha of land.
HEP	In the 1960s, Spain generated 70 per cent of its power from HEP but this has fallen to 20 per cent as newer technologies have been developed.
Water supply	20 per cent more dams have been built to address water scarcity – some regions and metropolitan areas lack adequate water supplies.
Flood control	Floods are common in Spain. Flood control dams are the main component of the flood control strategy. Several are currently being built – 13 in the Segura basin alone.
Recreation	Demand for recreation facilities in existing reservoirs is increasing due to hot, dry summers but there are concerns over maintaining water quality.

c) Reservoir capacity and reserves of water for consumer use

Drainage basin	Total capacity (hectometres – hm³)	Actual stored water as percentage of capacity, January 2002	10-year average of stored water as percentage of capacity
Duero	2780	35	60
Tagus	5710	45	46
Guadiana	8820	68	48
Guadalquivir	8660	70	48
Segura	1070	24	22
Jucar	3210	19	27
Ebro	3170	49	70
Combined east-flowing basins	26 560	60	50
Combined west-flowing basins	9100	34	47

ITEM 3

Water transfers from the River Ebro

a) The Ebro Water Transfer Plan

Map of proposed Ebro Water Transfer Scheme

The Ebro Water Transfer Plan involves building 120 dams and a 700km-long canal to transfer 1000 cubic hectometres (hm³) of water per year from the Ebro river basin in the north to the south-eastern Mediterranean coast. The estimated cost of the scheme is 20 billion Euros. Except for the 200hm³ to be used for supplying fresh water to the urban area of Barcelona, the remaining transferred water will be reserved for the agricultural areas that have 'irrigation rights'. Originally, the Government undertook to ensure that the transfer would not result in expansion of irrigated agriculture which currently consumes 80 per cent of Spain's water supply. An ecotax was also to be levied on transferred water to compensate donor regions. Currently, there are some doubts about both promises.

The government insists the plan will eliminate the historic imbalance between northern Spain, where water is considered plentiful, and the often parched south-east.

b) Proposed water transfers across eastern catchments in Spain

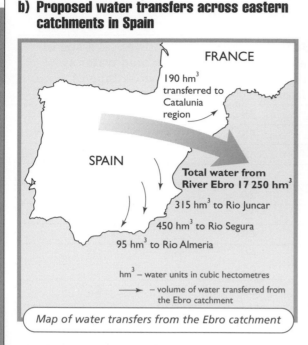

Map of water transfers from the Ebro catchment

c) The Ebro River delta

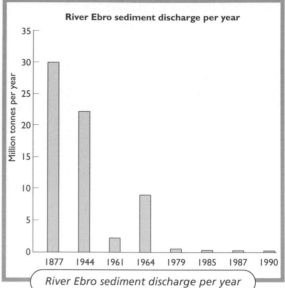

River Ebro sediment discharge per year

Since the mid-twentieth century, the Ebro river has been dammed in several places. This has reduced by 95 per cent the amount of silt flowing into the Ebro delta, a vast area of 32 000ha ideal for agriculture and the most important agricultural region of Catalonia. The delta is a huge nature reserve known for its biodiversity with a variety of different ecosystems, including sand dunes, reed beds and salt marshes. It is of international importance for birds, especially those migrating between northern Africa and western Europe. Flocks of flamingoes are one of the reserve's main attractions.

Approximately 47 per cent of the delta lies only 50cm above sea level with several cultivated areas lying below the level of the sea. There are annual losses of 15m on some of the delta coastline and, were the Ebro to lose even more of its flow, there is a danger that the delta will disappear within 50 years. The government has already spent 100 million Euros on replenishment schemes.

The delta faces eutrophication of lagoons through the use of fertilisers and pesticides, and human trampling of the delicate environment during the summer season. The current phenomena of coastal erosion and subsidence (due to soil compaction) are likely to increase and may be reinforced by sea-level rise in the medium term.

About 50 000 people live in the area, 1500 inside the delta and the remaining 48 500 on its inland edge, and 80 per cent of the land is agricultural or urbanised. Rice is grown over 65 per cent of the area. Salt mining is also an important part of the economy.

ITEM 4

Agriculture in Eastern Spain

	Total farmed area (ha)	Percentage of farmed area that is irrigated	Area under crops (ha)	Percentage of area under crops that is irrigated	Area under fruit (ha)	Percentage of area under fruit that is irrigated
Almeria	158	33	83	37	56	19
Andalucia	3544	22	1840	23	230	36
Aragon	1719	28	1516	21	111	34
Catalonia	816	39	514	27	129	52
Murcia	440	45	208	32	166	47
Valencia	620	33	114	39	309	67

Water in Almeria currently comes from over-exploitation of underground aquifers, 90 per cent of it used for intensive agriculture. It is estimated that 50–70 per cent of the 25 000ha under plastic is illegal and not registered for irrigation water.

There is little water regulation, accusations of mismanagement of water and apparent illegal water extraction. In Almeria and Murcia in particular, water is used for tourist development and golf courses, as well as for irrigated fruit growing for northern European markets.

ITEM 5
The Ebro controversy

a) The Ebro controversy

Agriculture

- The rainy season and growing season do not coincide so water-demanding crops like maize and potatoes are inappropriate.
- If water was not subsidised, 90 per cent of the land could not be profitably farmed.
- More irrigation will encourage the current inefficient agricultural system, using much water and fertiliser and relying on a large, and often illegal, immigrant workforce.
- In areas where water is expensive because it comes from desalination plants, farming is an efficient consumer.

Irrigation

- Irrigation water costs 1 per cent of the cost of water used in cities and industries.
- Taxpayers are subsidising the irrigation of crops that are already being subsidised by the EU and are often produced in excessive quantities.

Economic development

- Diversion of water will increase inequalities by accelerating the impoverishment and desertion of inland areas and encouraging uncontrolled development on the Mediterranean coast.
- Diverting water from the River Ebro will mortgage the present and future development of Aragon. The water will not be available for future developments there.
- Villages in Aragon will be flooded.
- People in regions such as northern Aragon and Catalonia, through which the Ebro flows, believe it is simply a means for the government to channel money towards Spain's powerful construction and electrical sectors. They claim it is more like a national cement plan rather than a national water plan.
- There has been a huge underestimate of the costs of the scheme.

Climate

- The Ebro basin has not had a water surplus for the past 25 years and projections are for a 15 per cent fall in rain owing to climatic change.
- Spain has more dam-created lakes compared to total landmass than any other country, but most are usually less than 10 per cent full.
- Five per cent of the Ebro flow diverted will be lost by evaporation from the canals and reservoirs.

Conservation

- A cheaper solution would be to invest 400 Euros per household in Catalonia for electrical and water appliances that can save and re-use water.
- Leaking pipes is a big problem, for example in Zaragoza where there is no difference between daytime or night-time consumption.

- Many farmers pay for water according to the area of land they need to irrigate so leakage does not matter to them.
- There have been no national studies of groundwater resources, water saving or consumption efficiency.
- Evidence from other areas suggests that diverting water, instead of satisfying demand, generates larger sustainable demands.
- The allowed ecological flow ($100m^3/s$) prescribed by the plan is not enough to guarantee the maintenance of the Ebro delta.
- The government has not considered any of the existing alternative solutions, such as recycling, increasing efficiency, or desalination.

b) European Parliament passes an important amendment against the Spanish Hydrological Plan

On 15 November, the European parliament approved a statement that expressed the parliament's grave worries about the development of unsustainable water management schemes and infrastructures by the Spanish government. 'The European parliament is deeply worried about the Spanish National Hydrological Plan as it does not address the issue of sustainable water use through pricing mechanisms and other water conservation measures.'

15 November 2001

c)

The Spanish government is defending its hydrological scheme. It claims that it is not catering for increased consumption but simply trying to meet the existing demand for water. Other plans would have a dramatic effect on Portugal by diverting rivers which flow across the border or would involve piping water from the River Rhone in France.

The government expects the EU to pay 40 per cent of the costs of the scheme, but the EU is under pressure, from local and environmental groups, to reject the plan as it violates EU habitat and wild birds directives. However, the government insists that the plan will solve Spain's water problem for good.

11 The Cairngorms – human impact in a wilderness environment

a

i) Use Item 1 to draw a graph to show the pattern of employment in Scotland and the Highlands and Islands Enterprise Area.   **8 marks**

ii) Suggest reasons for the differences in the employment pattern between Scotland and the Highlands region.   **8 marks**

b To what extent has tourism developed as a result of

i) physical factors   **6 marks**

ii) human factors?   **6 marks**

c Discuss the effects of the skiing industry on the physical and economic environment of the Cairngorms.   **10 marks**

d Assess the impacts of the building of the funicular on Cairngorm.   **12 marks**

Total 50 marks

Introduction

The 67 000 hectares of the Cairngorm Mountains, a national nature reserve, are situated in the Central Highlands of Scotland, some 200km north of Edinburgh and Glasgow. It is the largest area of highland over 900m in Britain and the least modified by humans. In this sub-arctic climate there is a range of significant and fragile habitats and species. Dramatic glaciated glens and lochs separate high passes and wilderness plateaux. Caledonian pinewood forests in the lower glens, moorland on the lower slopes and sub-alpine to alpine habitats on the summits host bird species such as Scottish crossbill and dotterel.

The mountains are managed by Scottish National Heritage to maintain the diversity and integrity of the natural heritage of the Cairngorms; to raise visitor awareness of the effects of disturbance and damage on sensitive habitats; and to develop the educational and interpretative potential of the area.

Aviemore is a skiing centre and summer holiday resort, and the largest settlement in the Cairngorm region. Three SSSIs lie immediately adjacent to the Cairngorm skiing area. The Cairngorms wilderness environment attracts 6000 visitors a day for winter skiing and mountaineering, and summer trekking.

Cairngorm plateau, Scotland

ITEM 1

a) Map of north and central Scotland

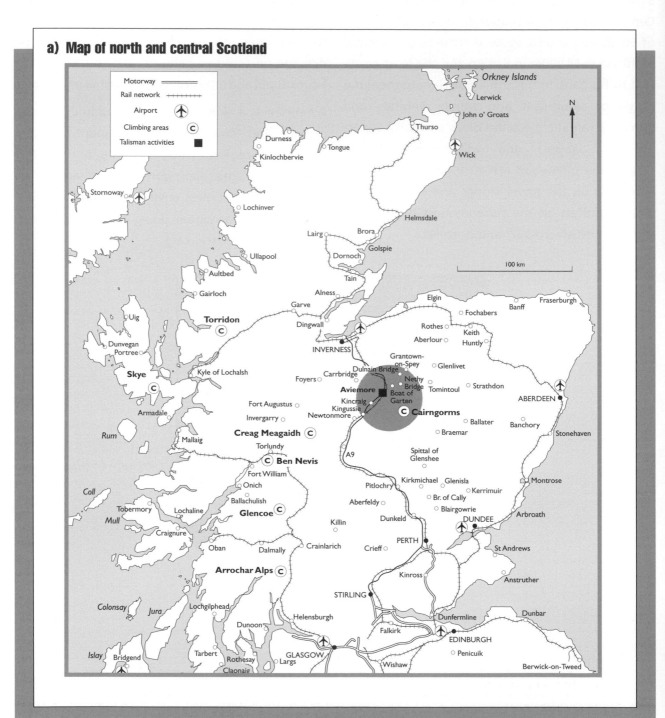

b) Map of northern Scotland to show Highlands and Islands Enterprise Area

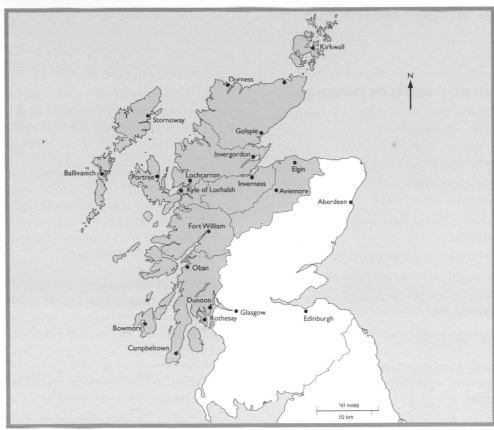

c) Pattern of employment in Highlands and Islands Enterprise Area, 1997

Employment categories	Scotland, percentage of workforce	Highlands and Islands Enterprise Area, percentage of workforce
Agriculture, forestry and fishing, energy, water	6.8	8.4
Manufacturing	16.2	10.1
Tourism and distribution industries	22.5	29.6
Construction, transport and communications	11.7	12.9
Banking and financial services	15.6	9.5
Public administration, education and health	27.2	29.5

ITEM 2
Tourism in the Cairngorms

a) Tourist attractions in the Cairngorms

- Rothiemurchus Visitor Centre – nature trails and guided walks
- Loch Garten Nature Reserve – nesting ospreys and lookout hide
- Abernethy Reserve – Scotland's finest Caledonian pine forest
- Glenmore Forest Park – designated for its wildlife
- Loch Morlich – watersports centre
- Loch an Eilein – historic centre
- Glenmore Lodge – Scotland's premier outdoor centre
- Mountain biking, horse riding and canoe hire
- Wildlife interpretative centre

b) Aviemore

Aviemore is a skiing centre and summer holiday resort, and the largest settlement in the Cairngorm region. The average length of stay for any overnight visitor is 1.2 nights. The Aviemore Partnership is trying to encourage visitors to stay for longer and make Aviemore their holiday base.

Local tourism-related employment in Aviemore amounts to 2500 jobs – well over half the total for the whole Cairngorm region. It is important for the survival of local businesses that tourists visit in both summer and winter seasons.

About half of all visitors are happy simply to be transported to the top of Cairngorm mountain, drink coffee and come back. Only 14 per cent wish to go for a walk of more than an hour. For 40 years, tourists and skiers were able to use a chairlift to access the top of Cairngorm itself. The chairlift has now been closed and a new, and contentious, funicular constructed in its place.

c) Skiing in Scotland

Skiing has become a major sporting and recreational activity in Scotland. Skiing developments are located on high ground in remote rural areas. Government policies for rural areas are based not only on sound stewardship of the natural and built heritage but also on the importance of creating the conditions for a healthy and growing rural economy. Skiing has an important role in bringing employment and other economic benefits to rural areas and, in so doing, provides a valuable contribution to the rural economy.

Government policy

Skiing in Scotland

A valuable industry

Controversial land use

Skiing in Scotland has considerable significance in the local economy of the Highlands region of Scotland. There are now over 70 chairlifts and tows available at five centres. During the period 1993–8 around 0.1 million trips each year were taken in Scotland purely for the purpose of skiing. Local expenditure on these trips amounted to around £16 million per annum. This brings valuable income when activity in other employment sectors (i.e. farming and summer tourism) is slack. Over 1000 jobs, including part-time and seasonal employment, are estimated to be supported by skiing alone.

As with all aspects of the tourist industry, skiing in Scotland is vulnerable to trends in employment and disposable income, but also to fashions in skiing location.

There are large year-to-year fluctuations in the number of Scottish ski days, but overall between the 1983–4 and 1998–9 seasons there was a small downwards trend.

Skiing is a controversial land use because it conflicts with conservation and other recreational interests. Chair lifts and associated buildings are unsightly; snow fences (designed to catch snow) produce an unnatural snow-line and changes to the flora as well being a barrier to hillwalkers. Reseeding of damaged areas leads to changes in vegetation structure and bird life. The Cairngorm Ski area lies in the largest privately owned bird reserve in Europe.

Soil erosion due to vegetation change and ski compaction can lead to flooding downslope. Out-of-season ski facilities encourage tourist access to sensitive areas with consequent trampling, erosion, litter and disturbance of breeding birds.

ITEM 3
The Cairngorm funicular

a) The Cairngorm funicular

The Funicular Railway Project involved replacing the prominent pylons of the outdated main chairlift system, with a low-profile funicular railway, together with new buildings incorporating an environmental interpretation centre. The funicular railway has been designed to minimise environmental impact, with use of advanced construction techniques.

To eliminate the skyline effect of the former chairlift, the uppermost section of the funicular runs underground in a tunnel. The replacement buildings use natural materials to blend with the landscape.

An estimated 200 000 visitors are expected each year for the funicular, compared with 50 000 for the previous chairlift, making the funicular an essential element of year-round tourism in the region. It will also provide for a greater variety of visitors such as day trips, families, coach tours, disabled visitors and educational and corporate groups.

b) View of the Cairngorm funicular

c) Benefits of the funicular

- Cairngorm is not a virgin site. The funicular is not an expansion but a replacement that will help to ensure the future success and further protection of Cairngorm mountain.

- The environmental damage of decades cannot be remedied without the complete removal of the ski facilities. The unsightly old chairlift has been dismantled and the middle station buildings demolished.

- A programme of environmental repair will be ongoing and subject to the most rigorous environmental audit to protect the surrounding landscape.

- The funicular and restaurant will work as a closed system; visitors will not be allowed out of the Ptarmigan building at the upper end of the railway and walkers will not be allowed in. Users will not be able to roam into sensitive areas of the Cairngorms, as did summer chairlift users (of 55 000 summer passengers, 10 per cent walked up to the plateau summit).

- Detailed visitor monitoring systems will be implemented. A high-quality interpretation centre within the complex will increase environmental awareness of visitors and promote environmental education.

- Visual intrusion of existing chairlift pylons will be replaced by the low-profile funicular route.

- Increased numbers of year-round visitors (insulated, for the first time, from sensitive areas) will create major added value for Highland tourism.

- Skier satisfaction and use will increase due to enhanced facilities.

d) Objections to the funicular project

The funicular will irrevocably damage the area's image of 'wildness and getting away from it all' – how can an area be wild if your granny can jump on a train that will take her right into the remotest interior?

The plan is for 165 000 visitors per year to use the funicular, which will run all year. However, because of the weather, the restaurant will provide a view for fewer than 40 days per year. Is this realistic?

Visitors will not be allowed to leave the restaurant area but this is only 150m down from the top – do they really think this will work?

Why are we increasing access to the mountain when, elsewhere, the National Trust for Scotland is actively removing access routes to the mountains to protect the wilderness?

Laying concrete and bulldozing new roads is not the way we should treat fragile mountain landscapes. Only those with no feel for mountains could do this.

Only people who walked up from the car park and have the proper gear to survive safely at the top in emergency should be allowed at the top of Cairngorm.

e) The Mountaineering Club of Scotland view

The Mountaineering Club of Scotland has consistently campaigned against the funicular railway on Cairngorm, believing that it would be economically unviable and environmentally destructive. The MCS wishes to protect the mountain environment and encourage appropriate and sustainable development in the long-term interests of visitors, residents and local businesses alike. The MCS recognises the importance of outdoor activities to the economic well-being of the local people. The MCS is closely involved in mountaineering activities in the area and its programmes attract significant numbers of people.

f) The Scottish National Heritage view

SNH objects to the size of the new Ptarmigan restaurant building at the summit and says that the development will increase numbers of visitors, all year round, posing a threat to the fragile environment.

SNH dislikes the 'closed system' idea because:

- it is discriminatory and should be a reserve measure;
- damage is minimal if visitors remain on paths (which have already been upgraded with public money);
- ranger guided walks from the Ptarmigan should be allowed.

a How does the physical environment of South Island influence the agriculture of the Canterbury Plains? **10 marks**

b i) Use the data in Item 1c to complete the graph Item 1e **6 marks**

ii) Summarise the importance of the Canterbury Plains to the agricultural economy of New Zealand. **8 marks**

c Outline the environmental reasons why the Opuha Dam was built. **8 marks**

d Assess the costs and benefits of the dam to
i) local farmers
ii) the wider community of the Canterbury region. **18 marks**

Total 50 marks

Introduction

South Island, New Zealand, has an oceanic temperate climate with relatively small variations in temperature (average: June 11°C, January 21°C). There are significant contrasts in rainfall across the Southern Alps; the west coast receives 5000mm of rain while the east coast has only 650mm.

The Canterbury Plains region is one of New Zealand's main economic units. High-tech businesses are concentrated in Christchurch, South Island's largest city but, surrounding the urban area, wide expanses of fertile, undulating land support productive agriculture. Timaru is the largest centre and port in the south of the region.

The Opuha River drains eastwards in the rain shadow region from the Southern Alps to Timaru and the Pacific Ocean. The Opuha Dam which opened in 1998 is primarily a source for an irrigation scheme to provide water in the unreliable rainfall conditions of the Timaru region.

ITEM 1
Agriculture in the Canterbury region

a)

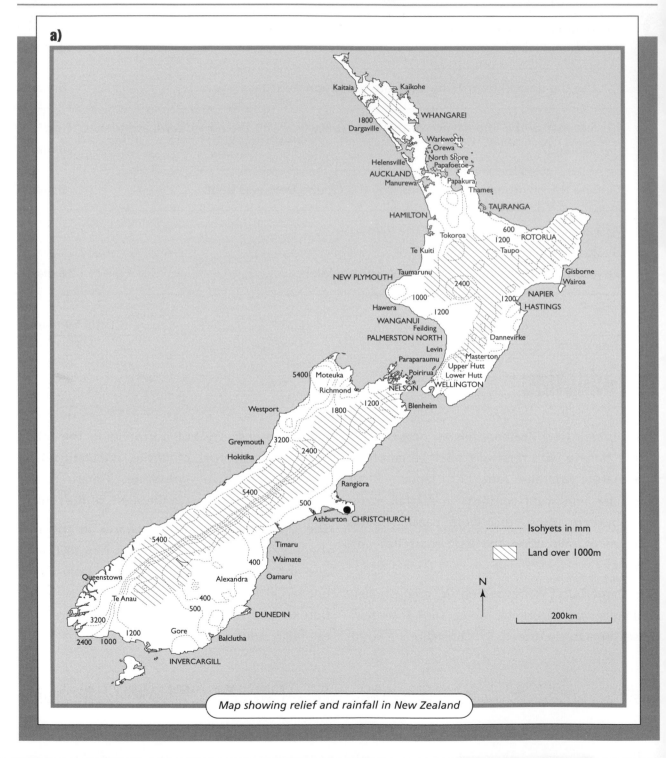

Map showing relief and rainfall in New Zealand

Isohyets in mm

Land over 1000m

200km

b) Agriculture in the Canterbury region

Farming dominates the economy in the productive Canterbury Plains with a mixture of intensive mixed farming on the plains and downlands, and horticulture near urban areas.

The Canterbury Plains is the largest flat, fertile plain in New Zealand and the moderate climate is well suited to arable, livestock and horticultural farming. Most soils are shallow on free-draining gravels and have a low moisture storage capacity. Although they are prone to drought, irrigation can increase their productivity.

Agriculture and its wholesale and retail industries are an important part of the economy. The Canterbury Plains have considerable potential for further intensification through double cropping and with value added through food processing.

Agriculture in the Canterbury Plains is changing as the Asian market opens up for high-value food imports. Recently many farmers have converted from sheep and beef production to dairy farming and specialist horticultural crops. The growth of vegetable processing has been particularly important.

c) Agricultural land use in New Zealand

Data for the percentage of total area for each land use			
Region	Percentage of grazing, arable and fodder	Percentage of horticulture	Percentage of other agricultural land use
North Island			
Northland	4.7	6.8	7.3
Auckland	1.9	–	2.2
Waikato	9.6	9.2	8.9
Bay of Plenty	1.8	4.8	3.4
Gisborne	2.7	5.8	4.0
Hawkes Bay	5.7	13.2	5.9
Taranaki	3.1	–	6.3
Manawatu–Wanganui	9.9	6.6	10.5
Wellington	3.2	3.8	4.5
South Island			
Tasman and Nelson	1.1	8.5	3.3
Marlborough and Chatham Islands	4.3	11.4	10.0
West Coast	1.3	1.0	3.2
Canterbury	24.0	17.8	18.2
Otago	18.2	2.1	7.9
Southland	8.3	1.0	4.5

d) Map of regions of New Zealand

e) Graph of land use in New Zealand

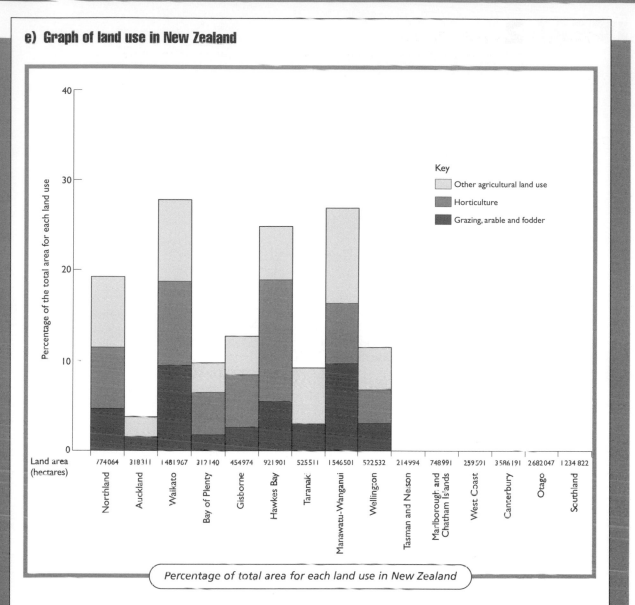

Percentage of total area for each land use in New Zealand

f) Employment change in selected industries in the Canterbury region, 2000–1

	Percentage change
Agriculture	+7.0
Manufacturing	−0.5
Transport and infrastructure	−4.4
Wholesale and retail trade	+12.0
Finance and services	+3.0

ITEM 2

a) Map to show location of the Opuha Dam

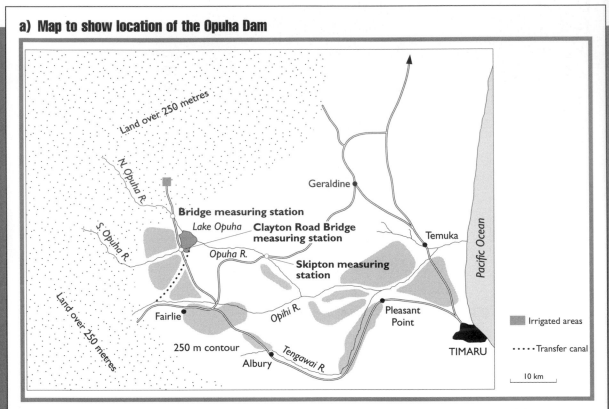

b) The Opuha Dam project

The Opuha Dam was opened in 1998 primarily to support an irrigation scheme for farmers in an area of unreliable summer rainfall.

The irrigation scheme supplies fairly small-scale water projects between 4 and 500ha. There is potential to supply 32 000ha of land but the decision has been made to aim for 16 000ha in order to provide a 98 per cent guarantee of supply to farmers from the South Canterbury and Levels Plain irrigation societies.

Agriculture has changed as a result of the

irrigation scheme. Conversion of farms from sheep to dairying, from animals to crops and from wheat to process crops such as peas will demand much greater labour input.

Key aspects of the scheme were:

- maintenance of river flow for ecological and conservation purposes
- water for domestic and industrial use
- water for irrigation
- production of HEP using all water passing through the dam
- control of erratic river discharges.

c)

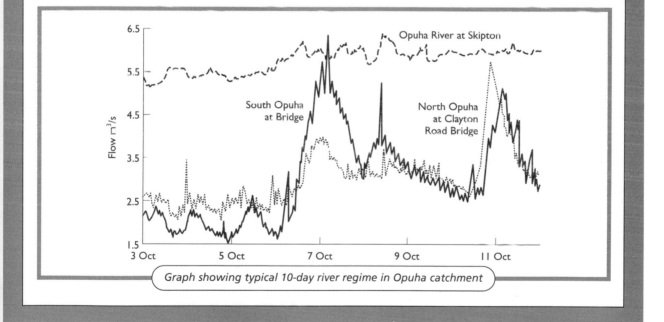

Opuha Dam, South Island, New Zealand

d) Hydrological impact of the Opuha Dam

Graph showing typical 10-day river regime in Opuha catchment

ITEM 3
Economic impact of the Opuha Dam

a) Changes in agriculture

	Net farm income per hectare (NZ$/ha)	
	Before the dam was built	After the dam was built
Sheep and cattle	300	500
Deer	500	700
Arable	900	1500
Crops for processing	–	Up to 15 000
Dairy	1200	2000

	Yields in tonnes/ha		
	Before dam was built	Anticipated yield with irrigation	After the dam was built
Wheat	6.5	8.0	9.0
Russet potatoes	38.0	55.0	65.0
Processed peas	3.6	7.0	5.5
Grass seed	1.2	1.4	1.7
Clover seed	0.2	0.4	0.4
Brassica seed	1.2	1.3	1.4

b) Potential benefits of the scheme

Sources in Timaru claim that the Opuha Dam will reverse recent economic decline. The Plains Irrigation Scheme alone (outside Timaru) is likely to inject NZ$16 million per year into the local economy, including 17 full-time jobs. Each full-time job may create seven more jobs in local infrastructure.

Land values have already risen as North Island farmers move into the area to benefit from the irrigated land.

Local fishing has benefited as the constant flow from the new dam keeps the river mouth open for salmon and has stabilised the habitat for trout, whitebait and eels. Fish production could double in five years, according to local anglers.

McCain Foods is one of a number of firms expanding their plant to cope with increased farm production. A spokesman said 'local potato production is likely to rise from the current 20 000 tonnes. We can sell all we get into the expanding chip market in South East Asia. We can promise high quality with the short time between harvesting and processing and can ship out of Timaru or air freight from Canterbury.'

ITEM 4
Tourism at Opuha Dam

a) Tourism at Opuha

Improvements in the local hotel trade began even during the construction phase, and employment in tourism, transport and electricity generation will provide an immense boost locally.

Tourism will be further boosted by news that the Opihi Transfer Canal to the reservoir will ensure the lake is full for 45 per cent of the year, allaying fears that in late summer mudflats would discourage visitors. However, Timaru District Council is still unwilling to allow camping, in order to maintain water quality.

b) Tourist use at Opuha

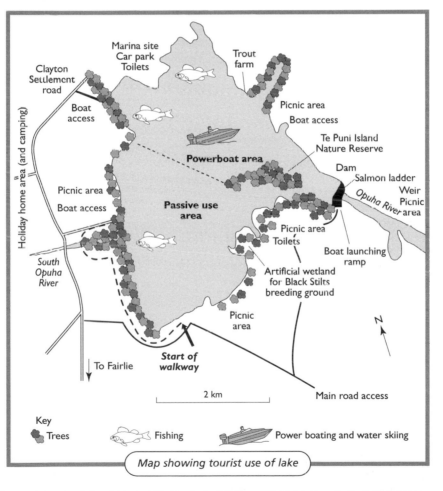

Map showing tourist use of lake

ITEM 5
The 1997 flood

a)

In 1997, the erratic river regimes which are a characteristic of this area proved too much for the construction engineers. During its construction, the earth dam at Opuha was dramatically breached after heavy rains caused the lake to overtop the unfinished dam.

Despite the disastrous flood, and after a legal battle concerning liabilities and NZ$1 million compensation, the Opuha Dam is now seen as a major economic asset to the South Canterbury region.

b) Disaster highlights environmental risks

On 5 February, heavy rains caused the lake to overtop the unfinished Opuha Dam. In the absence of an adequate concrete spillway, the water cascaded down the outer wall of the dam, rapidly cutting a gorge. One-third of the dam, 200 000 cubic metres of material, was lost and 2 million cubic metres of water emptied down the Opuha River overnight. Some 20 farms suffered major damage but there was no loss of human life. Trees were uprooted, bridges destroyed and 8000ha inundated. Paddocks were flooded, killing livestock and leaving them smothered in shingle. There was total fish loss for 12km downstream, combined with more extensive destruction of other aquatic fauna by sedimentation in the lower course. Eels (a key element of Maori cooking) were badly affected but, fortunately, most salmon stocks were at sea at the time. Effects on the food web also resulted in the decline in numbers of some bird species. However, scouring of pools also provided better habitats for returning salmon and cleared the river mouth to allow more spawning salmon upstream.

a Explain how the physical environment poses problems for the city of Venice. **10 marks**

b Suggest how the physical environment of Venice may change
  i) in the short term
  ii) in the long term **12 marks**

c i) Using the data in Item 1, draw a graph to show population change in the Venice region. **6 marks**

  ii) Suggest reasons for the pattern shown on the graph. **6 marks**

d To what extent are the problems facing Venice caused by the tourist industry? **8 marks**

e Identify possible strategies by which the civic authorities can ensure a sustainable future for Venice. **8 marks**

Total 50 marks

Mark scheme

a

Should identify:

- lagoon, sand foundations
- 120 islands lead to problems of space, stability, congestion
- problems of transport
- disposal of waste
- global warming – more erosion
- tides and saltwater – building damage
- 'closed' nature of lagoon – chemical composition – eutrophication

1–3 marks	Describes some aspects of the physical environment. Implied understanding that physical environment causes problems.
4–6 marks	Some simplistic explanation though not at all comprehensive. Some links recognised between physical environment and specific problems.
7–10 marks	Comprehensive explanation of a range of physical factors and problems. For highest marks, appreciates scale of problems posed by physical environment.

b

Short term	Long term

Short term

- continued deterioration of physical structures
- rising water levels/flooding
- declining species and rising turbidity
- reduction in speed limits therefore reduction in erosion

Long term

- removal of waste dumps
- sewage treatment
- education of farmers
- environmental controls on industry
- more breakwaters, beaches, dykes for coastal defences

6 marks for each section. Must clearly explain 'how'.

1–2 marks	One idea outlined about changing physical environment.
3–4 marks	More than one idea identified. Some explanation of the physical environment. Does not convincingly explain short/long-term change. Alternatively, gives very brief outline of a wider range of changes.
4–6 marks	At least two ideas well developed. Specifically recognises short/long-term change and explains 'how'.

c

i) 1 mark for accuracy of each line, and for correct labelling of axes. Must be appropriate choice of graph.

ii) Reasons for:
   decline in Venice city
   decline in population on islands
   growth then decline of population on mainland.

Outline reasons for each line – 3 × 1

Detailed reasoning for each line – 3 × 2

- tourists are oppressive – noise, congestion, litter
- rents rise as a result of tourism, also living costs
- damp, old buildings and high repair bills
- lack of vehicular access and egress
- stabilising population and increase in young population who are less interested in old city
- islands are inaccessible
- possible initial flight to suburbs
- employment elsewhere, not in tourism

d

1–2 marks	Describes one or two problems in Venice with implied reference to tourist industry.
3–5 marks	Identifies specific links between tourist industry and problems in Venice but does not recognise scale of impact.
6–8 marks	Addresses 'to what extent'. Concerns a range of problems and clearly examines their causes in relation to different levels of tourism activity. Balanced discussion.

e

- acting as a unitary body
- stemming population exodus to create population balance
- revising image of Venice as a museum
- investment in public services, proper sewage systems
- enforcing strict environmental controls
- providing flood protection for worst-hit areas
- encouraging growth of non-tourist services, especially for young people

1–2 marks	Lists one or two strategies. No reference to 'sustainable'.
3–5 marks	Identifies some strategies with implied links to 'sustainable'.
6–8 marks	Recognises a range of strategies appropriate for civic authorities. May comment on relative importance of each. Specific comments on how each strategy will lead to sustainability.

Specimen answer

a The physical environment poses several problems for Venice. *This opening sentence is unnecessary but it may help candidates to get started.* The small size of the island, 50 × 150km results in very high visitor densities and crowded streets all year round. *Uses detail from the Introduction.* This spoils the enjoyment of the unique architecture and atmosphere. Lack of space is a problem for new developments and modernisation. Efficient communications in Venice is problematic with so many crowded waterways in which a variety of freight and passenger traffic pollutes the water. *Specific problems linked to physical environment.*

Because Venice is built on a sand bar, foundations of buildings are increasingly insecure. High tides create strong currents which are eroding the mud flats and salt marsh which have been protecting the islands and Venice city. Tides also affect foundations and polluted water increases chemical weathering of stonework. Demand for fresh water led to pumping out of groundwater and, consequently, some subsidence of Venice's buildings. This is a particularly serious problem because the impact of rising water is putting more pressure on foundations. *Scale of problem recognised here. Good combination of detail to emphasise the scale and temporal aspect of the answer.*

Global warming is also leading to more regular flooding of Venice's famous landmarks. This leads to further weathering and destruction of the streets and architecture.

The physical characteristics of Laguna Veneta and the long, narrow barrier to the Adriatic Sea also cause problems for Venice city itself. The narrow entrances to the lagoon limit the removal of polluted water and the build up of high concentrations of pollutants increase chemical weathering and, in turn, erosion of Venice's buildings which border the lagoon.

Because Venice is an island, a major problem for the authorities is the disposal of waste. *Recognises scale here.* It could be argued that as fewer people live in Venice, the problem is less acute, but the large number of visitors will produce vast quantities of biodegradable and non-biodegradable waste. In the days of increasing environmental awareness, dumping untreated waste into Laguna Veneta is not a sustainable option.

Range of specific problems linked to physical environment. Both human and physical problems identified. Answer could be better structured (e.g. consider major problems first).

b

i) Short-term impact is taken to mean less than a decade. *Fair comment to identify what is taken as short term. Some students will be more specific, relating this to the economist's concept that it is a period too short for changes in infrastructure.* Over the next few years, Venice's sandstone buildings will continue to erode dramatically and the detailed stonework carvings will disintegrate. Flooding will become progressively more severe with rising sea levels and the force of high tides through deepened channels at the entrance to the lagoon. The lagoon will become progressively more polluted until there are strict, enforced pollution controls. Species diversity and water turbidity will increase. *Could have used data here from resources.* More buildings will deteriorate as costs of maintaining them rise and people leave the city. A short-term improvement could be made by reducing speed limits for boats along the canals. This would reduce the wash from boats, reduce wave impact on waterfront buildings and slow down the rate of destruction. *Good depth of explanation for high level answer.*

c

i) The small rise in island population between 1960 and 1970 may be a result of the attraction of living in an isolated location but with some improvement in transport and services. However, the overall decline in the population of Venice and the islands is probably a result of poor environmental quality and inundation by tourists each day. Expensive maintenance of buildings and the inconvenience of increasingly frequent flooding encourages people to leave. Lack of jobs and facilities for young people in Venice also encourages out-migration.

The increase in mainland population between 1960 and 1975 was probably a result of industrial development and new jobs at Porto Marghera. However, gradual decline since 1975 may reflect increasing environmental pressure – air and water pollution, less space and high cost of living near the lagoon.

Comments on each line of the graph in turn and recognises change within each location. Some detailed reasoning offered. Plausible suggestions with a clear focus on pattern.

ii) In the long term, the physical environment may change for the better as more and more people are concerned to protect Venice. Farmers on the mainland may be forced to control their fertiliser waste which would reduce agricultural pollution in the lagoon. The building of a sewage treatment works for Venice could also improve lagoon water and reduce unpleasant smells substantially.

However, rising sea levels will present a long-term impact without some form of flood protection. The large-scale engineering works needed for this will take some time to come into effect.

Three ideas offered here although link to long term not always clear. Comments are quite brief but does explain 'how'.

d Some *Hints at evaluation* of the problems of Venice are unrelated to the tourist industry. Rising sea levels are a global problem and affect all low-lying coastal areas. Venice is of particular historic importance so the result of flooding is of international concern. Pollution in the lagoon is not really caused by tourists but there is a risk that pollutants from visiting boats do contribute to the pollution cocktail. Farmers and heavy industry contribute more pollution than visiting passenger ships.

By considering problems unrelated to the tourist industry the answer begins to address 'to what extent'.

Wave erosion by boats in the canals is partly caused by tourist gondolas which ferry people around the canals in very large numbers. Tourists also contribute to the accumulation of waste in Venice, particularly with the rapid increase in the number of fast-food outlets. They therefore accentuate the waste disposal problem for the civil authorities. Erosion of buildings and statues occurs more rapidly as a result of tourists who cannot resist the temptation to touch the statues. Many visitors wish Venice to remain as a museum piece and do not support the modernisation of the city. The civic authorities are concerned about the lack of money generated by tourists within Venice – income which could be used to pay for environmental repairs. The problems of out-migration may also be accentuated by tourists as local people feel pressured by the large numbers of visitors each day.

It is also true, however, that many of Venice's problems would exist regardless of tourism. Expensive repairs, lack of space to modernise and lack of facilities for young people are concerns for local people rather than the tourist industry.

Good range of problems treated systematically. Some specific assessment of 'extent'.

e There are a number of strategies which could lead to a sustainable future for Venice but some depend on national rather than civic action. The civic authorities need to act as a unitary body to ensure that all policies are complimentary. Improving the quality of the environment through improving public services will encourage people to return to live in Venice although there will be a limit to the number of people the island can support. Another strategy would be to control the speed of boats on the canals and perhaps limit their number to avoid congestion on waterways.

The civic authorities could limit the number of tourists entering Venice but they should be aware that the tourists who create the pressures could contribute financially to the sustainable future of the city.

Tends towards an implied answer in places. Recognises a range of strategies and notes which would be appropriate for civic authorities but links to 'sustainable' could be much more explicit. This is a rather short answer for 8 marks.

In general, this is a good answer although the student could make more specific references to the resources. The text is coherent and the ideas are clearly explained but some of the answers could be better organised. The words from the questions are used in each answer. This helps to keep a focus on the precise question set. There is no need for an introduction to each section.

a

i) Use the outline map (Item 1a) of the Environment Agency regions of England and Wales to represent the data in Item 1b for household water consumption.   **6 marks**

ii) Using the resources in Item 1 to help you, describe and suggest reasons for the distribution you have mapped.   **4 marks**

b

i) Use a second copy of outline map Item 1a to represent data in Item 1d for water abstraction.

6 marks

ii) How is the pattern of water abstraction affected by the physical environment?   **6 marks**

c

Suggest how water demand appears to be influenced by changes in
i) land use   **6 marks**

ii) population distribution   **4 marks**

iii) changes in local economies   **6 marks**

d

In what ways might the changing patterns of population, industry and land use have an impact on water management in England and Wales?   **12 marks**

Total 50 marks

Mark scheme

a

i) Appropriate style of map and accurate key – 2 marks
Could be a choropleth map or a located bar graph.
Accuracy of plotting data – 8 × ½ marks

ii) Broad pattern of greatest consumption in S and E. Least in N and W.
Reasons would include:

- lower rainfall in SE, therefore greater demand on water for gardens (e.g. sprinklers)
- greater wealth in SE, more household appliances (e.g. dishwashers)
- more homes with more than one bathroom
- more people living in S and E in single-person/small households, therefore more consumption. People still have to water the garden/wash the car/clean the house
- south and east get more tourists in summer who boost water consumption

| 1–2 marks | Basic description offered. S and E / N and W contrasts. Some simplistic reasoning for 3 marks. |
| 3–4 marks | Detailed specific description. Uses data from map and tables to develop well-reasoned points. |

b

i) Appropriate style of map and accurate key – 2 marks

Could be a choropleth map or a located bar graph. Accuracy of plotting – 8 × ½ marks

ii) For full marks, students are expected to have some outline understanding of the different physical environments across England and Wales (e.g. outline references to temperature, geology, location of uplands).

- Greatest water abstraction in SE and Midlands. Least in SW and W. Reason – W relies less on water abstraction because reservoirs maintained by high rainfall and low evaporation rates. In S very low rainfall totals in summer and winter but high demand from households and commercial uses such as offices.
- Geology in SE suitable for aquifers (e.g. artesian wells in chalk).

- High levels of abstraction in East Anglia because low rainfall. Uplands of N and W provide opportunity/locations for reservoir construction. Less suitable geology in S and E.
- Fewer sites in S and E therefore more reliant on water abstraction from major rivers and aquifers.

1–2 marks	Basic description offered. Recognises difference between N / W / SW and Midlands, SE.
3–4 marks	Detailed description. Some use of data. Reasoning focuses on rainfall only. Relevant use of data for 4 marks.
5–6 marks	Thorough description. Recognises strength of contrasts using data. Explanation uses seasonal rainfall data but also some reference to geology, topography, use of reservoirs in N and W rather than major rivers and aquifers.

c

i) Land use

- more houses built; substantial increase in land becoming urban in S and E and East Anglia
- intensification of arable farming may require irrigation; livestock need water for milking parlours
- leisure uses – golf courses use more water

1–2 marks	Simplistic view of one change in land use. Change from rural to urban. 'More houses therefore more water consumed.'
3–4 marks	Comment on more than one land use. Some use of data. Brief reference to 'change'. Outline suggestions only.
5–6 marks	Discusses at least two land uses and clearly addresses change. Plausible suggestions based on sensible geographical principles. Uses evidence from resources Items 1 and 3.

ii) Population distribution

- greatest net migration to SE and East Anglia therefore more demand for water there (Item 1b)
- more affluent people move and therefore more water consumed
- increase in land under urban use as a result of changing population distribution
- substantial in-migration to SW matched by loss of rural land to urban; few industrial sites to be redeveloped for housing

| 1–2 marks | Notes population distribution – more people, more water. Some outline evidence from resources for 3 marks. |
| 3–4 marks | Specifically discusses population *change*. Detailed use of resources to link population distribution to increased water demand. |

iii) Change in local economies

■ decline in heavy industries in N and W reduces demand for water; ratio of new investment to manufacturing output; Item 4c reflects decline in heavy industry and replacement by new manufacturing which is low input/high tech, therefore needs less water

■ footloose industries, office/service industries require less water

■ growth of leisure industries increases demand for water

1–2 marks	Brief description only of declining heavy manufacturing therefore less water.
3–4 marks	Some valid suggestions but very generalised references to industrial change.
5–6 marks	Specific references to industrial change and demand for water. Includes some regional detail.

d

■ generally increased demand although domestic/industrial balance has changed

■ greater imbalance in regional demand for water, therefore management requires inter-basin transfers

■ issues raised about cost, environmental desirability of inter-basin transfers

■ potential sources of reservoir water in N and W raise local issues of flooding valleys

■ more emphasis on efficient use of water and recycling water especially through increased pressure on water resources in S and E

■ regional imbalance exaggerated by problems of climatic change and seasonal drought in S and E becoming more common

■ out-migration from London and industrial decline in the city therefore rising groundwater table

- issues for Anglian Water – large net population increase yet no obvious water transfer scheme. Risks from rising sea levels affecting groundwater quality; management focus on water conservation and efficiency
- industry – decline of heavy industry, therefore fewer pollutants, therefore cleaner rivers, therefore more water available for household abstraction
- land use increasingly urban, therefore effects of impermeable surfaces, less infiltration, reduced groundwater recharge

1–4 marks	Generalised comments which describe 'more demand therefore greater impact'. Limited use of resources.
5–8 marks	Recognises how change impacts on water management. Lifts ideas from resources but does not develop clear explanation or regional detail.
9–12 marks	Examines specific management issues related to changing water demand. Range and depth expected. Notes specific impacts and regional differences.

Specimen answer

a

ii) The greatest household consumption is in the South and East of Britain, and least in the North West.

This may be because higher incomes in the South mean homes have more consumer goods (e.g. dishwashers) which use more water. Lower incomes in the North and West mean fewer such goods. The South East is generally warmer in summer and so there is more demand for garden sprinklers, paddling pools, etc.

A sound answer but it does not fully exploit the resources. It could have considered variations in consumption in more detail (e.g. the higher prevalence of single homes in the South East).

b

ii) Greatest water abstraction takes place in the South East and the Midlands. The South West and West depend less on abstraction because reservoirs and rivers are easily filled by high rainfall. High land in the North means that valleys can be flooded for reservoirs.

In the South in summer there is high demand from leisure activities but low rainfall and so decreased river flows, and this creates a shortage of water. Water companies need to get water from rivers and also use underground water resources.

Most key influences on pattern are identified, although there is little evidence of the use of specific resources. Quoting detail flags up that resources have been used (e.g. 'Greatest water abstraction takes place in the South East and the West Midlands ... Over 1000 million litres per day compared to 500 million litres').

There is room for expansion and detail here. Why is land in the North (and West) suitable for reservoirs? Can the Pennines be mentioned by name? What about geology? – chalk aquifers are important in the South. Can/does the South provide for all its needs by tapping aquifers?

c) i) Item 3 shows that 20 000 new houses were built in the South East in 1999 and 16 000 in East Anglia. Also that both regions have had a large change from rural to urban land use. Urban areas use more water.

If farming is intensified, say in agriculture, it is likely that irrigation will be used. This change in land use also requires more water.

In many areas near towns rural land is being turned into golf courses. These require plenty of sprinklers to keep the greens in good condition which means changing land use increases demand for water.

It would be helpful to be more explicit over the LINK between the geography of house building/urbanisation in England and Wales and demand – as shown on the map.

ii) The greatest net migration is in the South East and East Anglia and obviously more people require more water. It tends to be the most affluent people who move and they are also likely to have a higher water consumption. There are more people living in the South West now who will increase the demand for water. These are often retired people who enjoy leisure activities which could therefore increase demand for water even further.

A good answer, but there are missed opportunities. It does not exploit the detail from the resources.

iii) Old heavy industries used to use lots of water but as they have declined there is less demand. Newer industries which have replaced them are high tech and use much less water. There was a concentration of heavy industries in the North West and North East, both regions which attract government incentives to encourage more inward investment. As this investment is in modern industry, demand for water falls.

Demand is also falling in some of these areas as mineral exploitation winds down as a consequence of depletion or foreign imports. Increased efficiency – partly driven by the high cost of water – has also held back increases in consumption. It is important to do the extra distance to achieve the top marks – these cannot be attained with simple answers which are not developed.

d The changing patterns of land use, population and industry have major implications for water management in England and Wales. The South East will become increasingly short of water especially as global warming leads to rising temperatures and more frequent drought. This, in turn, means demand for water for gardens, etc. but less rain to recharge underground water supplies. The demand in the South East may be met through water transfer schemes from rivers in the North and West. Water transfer affects river channels and upsets hydrology and wildlife in rivers. It is also expensive. People need educating about water conservation in homes. Also leaking pipes in streets could be mended.

There will be opposition from people in the North and West if more reservoirs are built. Although they can be used for recreation there is still concern over flooding land and changing the character of the countryside, perhaps even flooding small villages as in Mardale in the Lake District. Even if reservoirs are enlarged, rural land and farmers' livelihoods are affected – just because of water shortages in the South.

One good change is that as heavy industry has declined there is less polluted water in the waterways and so more can be recycled for domestic and light industrial use. Water in London is now recycled seven times.

This is not specific in linking water shortage in the South East to water transfer schemes. Insufficient time seems to have been left here and this is a brief set of mixed ideas with some relevance to the question set. There are clear stimuli from the resources but they are not properly linked into a cogent argument. Regional differences are seen but not developed, exemplified or used in a coherent argument.

Preparation

To begin, take a few minutes to read through the resources and become familiar with the topic. Don't forget that the introduction is critical. It introduces key ideas about the topic in a very concise way.

Be prepared to think about each answer before you start to write. You will write more effectively if you know what you want to say and where you are going.

a **Outline the geographical characteristics needed for a successful container port.**

8 marks

The command word is 'outline' (i.e. give a brief, concise explanation of characteristics ... in this case which lead to a *successful* container port). Aim for two or three sentences on each point. The bullet points in the Introduction easily identify characteristics, but be careful – they must be *geographical*. Watch out for plagiarism – although it will get some marks, you will be a long way from a high-level answer. What geographical characteristics can you determine from the map?

b **i)** **Using Item 1a, draw a graph to compare the percentage change of container movements in the five largest ports in the UK from 1989 to 1999.** **6 marks**

Your graph may take any form, provided it serves its purpose. You must be able to *compare* the changes over time. Consider carefully what type of graph is most appropriate that you can draw quickly. It is not meant to be a work of art, but a means by which you can further analyse the data.

You will be marked on the accuracy of plotting plus correctly labelled axes.

ii) **Comment on the relative growth of container movements in these ports.**

8 marks

'Comment on' means just that. Make some sensible, logical, geographical statements about the growth of each port in turn, *relative to the others*.

Think about *rates* of change.

Don't forget that you have graphed percentage figures. High-level answers will recognise relative growth in terms of absolute figures ('000 container units) too.

For full marks you will need *at least* four developed comments with a specific focus on 'relative'. To achieve this you will need to include evidence – data – from Item 1a or your graph. You could consider annotating your graph as part of your answer but it would be more difficult to achieve full marks. You must make it very clear to the examiner that your comments are represented as annotations. Beware ... do not then repeat yourself in the text.

iii) **Suggest why new developments are planned at all the major container ports in the UK.** **6 marks**

'Suggest why' means that you can give some plausible reasoning based on your acquired geographical knowledge and logic.

Read the Introduction again to remind yourself of its contents. You have probably forgotten much of it while drawing the graph. What does it suggest about new developments?

There is more information in Items 1b and c. Think carefully about what the data means and why European ports are included in the list. If there are new economic developments of any kind, what effects might they have on the local or national economy?

c **What are the advantages of Dibden Bay for a new container port?** **6 marks**

This is a fairly straightforward question in which you have to identify the positive aspects of Dibden Bay using all the resources in Item 2. Do not be tempted to lift information from 'The Plan'. You need to summarise what is presented. Remember that the focus is on what makes Dibden potentially a good *container* port. Make sure that you make specific links to address the question.

d **The opponents of the Dibden Bay scheme make their case on ecological, economic and environmental grounds. Assess the validity of their arguments.** **14 marks**

This section is clearly very weighty with 14 marks to award. You need to move on to Items 3 and 4. Remember that you already know something about

■ what a successful container port requires
■ how the UK container ports have changed in recent years
■ what advantages Dibden has as a proposed site.

The command here is '*Assess the validity*'. You are unable to do that without weighing up *all* the arguments – positive as well as negative.

Make sure that you are familiar with the proposed plan in Item 2. Look carefully at the map in particular.

If you look intelligently at Item 3 you will recognise that a division of views has been made for you – into economic, ecological and environmental. Note also that some views are positive rather than negative.

The risk here is that you take each view in turn and assess it. However, you will run out of time and probably make the same points several times.

Instead, consider each *group* systematically. Consider the legitimacy of the views in the context of the geographical knowledge and understanding you have acquired through the A level course. Remember that you do not need to refer to *all* the views presented in Item 3.

In your planning you could organise your ideas in a table. Although this takes time, it might help you to write more fluently and make better assessments.

For example try this table for the environmental argument:

Opponents' views	Proponents' views
Large cranes spoil the landscape quality (you could refer here to photograph Item 2d).	Improvements to surface drainage and pollutants discharging into Southampton Water will raise water quality.
More dredging in the channel.	
More erosion and therefore impact on nearby coastal areas.	
Oil pollution.	
Industrial development would spoil the New Forest, a new National Park.	

Think about the groups with particular interests in the Dibden Bay proposal. Remember that the views which any one group holds depends on their background, role and responsibilities in the wider community. In your assessment you should be aware of any bias suggested by any particular group.

You will be marked on your ability to organise all these ideas. Make sure that you specifically 'assess the validity' of the views. You must clearly justify the opinions you reach concerning all the arguments.

And finally ... you could make a brief personal assessment on whether or not you think the scheme should go ahead.

Preparation

To begin, take a few minutes to read through the resources and become familiar with the topic. Don't forget that the introduction is critical. It introduces key ideas about the topic in a very concise way.

Be prepared to think about each answer before you start to write. You will write more effectively if you know what you want to say and where you are going.

a   **Describe and suggest reasons for the economic problems faced by the Ruhr valley in the 1980s.**   **12 marks**

Note 'suggest reasons for'. You should be giving plausible suggestions. You are not expected to have a detailed inside view of the Ruhr. Use your geographical knowledge and common sense here.

- What happens when manufacturing declines?
- Who is affected?
- What happens to the physical environment?
- Does that affect economic issues?
- Consider how the *types* of industry which dominated the Ruhr Valley would help your reasoning.
- How might the *global economy* have affected the Ruhr region?

Look carefully at Item 1

In 1a can you link the urban problems to economic issues?

In 1b identify the trends in employment structure. What economic problems do the changes suggest?

1c sets the Ruhr in the context of other German industrial regions. Do the contrasts shown here suggest further reasons for economic problems? What *data* could you use to support your reasoning?

Look at 1d at the scale of unemployment in the Ruhr compared with the German average. What are the links between unemployment and economic problems?

Also in 1d, GDP per capita is a measure of income. Look at these figures in relation to the average for Germany. Note the date – 1986. Does the situation in Oberhausen, Bottrop, Duisburg and Dortmund suggest any problems?

Remember that there are two elements to the answer. You need to *describe* and *explain*. For 12 marks, a substantial answer is expected. Remember that if you have thought about the resources and the question first, you will write more effectively.

b   **Summarise the challenges to planners and developers in the Emscher valley from**
     i)   the physical environment
     ii) the social and economic environment
     16 marks

N.B. If the marks are not specified for each part, there is some flexibility in your answer. It need not be a perfectly even treatment of i) and ii). However, you **must** consider each section as fully as possible.

i) the physical environment

Here you need to use further resources but don't forget what you already know about the topic.

Begin by listing aspects of the physical environment which could be seen as problematic/challenging for planners and developers. Item 1a lists some characteristics of the urban environment. Item 2a begins with a description of the Emscher valley landscape.

When you have a range of ideas, try to classify them; arrange them into groups, such as those below, so that you can explain concisely why they are a challenge to planners and developers.

Item 1a Characteristics of the physical environment	Surface subsidence, polluted waterways, slag heaps, tall unsightly buildings
Item 2a Emscher valley	Industrial buildings – coking plants, iron foundries, coal mines, soil contamination, abandoned factories

Classify aspects of the physical environment:	Land – subsidence
	              – rivers
	              – soil
	Buildings – size
	                  – deterioration

Why do you think it is challenging for planners and developers to deal with these aspects of the physical environment?

What is a challenge? Something which is not straightforward; may be unusual; may take some time. Think about the challenges listed here.

Use your knowledge from work you have done on industrial location/managing the environmental impact.

ii) the social and economic environment

Here you are building on knowledge identified in a). You should now be aware of economic problems. What social problems also result from economic decline and a poor-quality physical environment?

Use your geographical flair and logic. If you were living in the Emscher valley, what social problems would you face?

- unemployment
- less income circulating in the local economy, therefore recession in local shops and businesses
- de-skilling of local workforce
- lack of self-esteem, particularly for young people
- increase in crime, drug abuse, alcoholism
- out-migration of able, energetic young people

Why are these issues seen as challenges? Why are they difficult to solve?

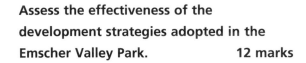

c **Assess the effectiveness of the development strategies adopted in the Emscher Valley Park.** **12 marks**

Begin by listing the strategies adopted. Use your knowledge from Item 2b. 2c will also suggest some ideas. Move on to Item 3. Don't forget to use the photograph in your assessment of 'effectiveness'.

Strategies

What was the basic principle? To use the industrial history positively and provide something sustainable for future generations.

Strategies	Assessment
■ water park along canal system ■ convert industrial buildings to cultural/museum use ■ encourage small-scale businesses ■ renovate decaying areas (e.g. Duisburg harbour) and create mixed, coherent neighbourhoods ■ improve environmental quality everywhere	Your personal opinion is valid here but it must be well argued and justified, using evidence from the resources and knowledge from your other geography units. Don't forget that things are not perfect. If you do not think a particular strategy is effective then say so, and justify your view.

d Examine the advantages and disadvantages of large environmental management schemes such as the Emscher valley.

10 marks

In this final section it is assumed that you know a good deal about

■ the industrial history of the Emscher valley
■ the origins of environmental problems
■ the impact of those problems on the physical, social and economic environment
■ some detail of strategies adopted to solve the problems.

Now you need to have an overview of the whole scale of the scheme. Don't forget to add Item 4 to the other resources you can use.

N.B. 'Examine' means look in detail.

Before you begin, organise your ideas such as those in the table on page 124. You may disagree with the list. As long as you explain clearly, your view of advantages and disadvantages is perfectly acceptable.

Advantages	Disadvantages
Improves environment for the future. Makes polluting industries contribute to the scheme. Encourages people to work together. Provides something for whole communities, not just those looking for work. Creates a sense of local pride, self-esteem. Values and integrates recent history of region rather than wipe it out.	Co-ordination of different groups of people/professional/interest groups is difficult. Limited life-span of project. Difficulty of maintaining impetus after project closes. Lacks emphasis on jobs. Does not always bring sufficient new money to circulate in the local economy. Huge cost.

In your answer organise your ideas and clearly explain each point. You could make a concluding comment on the balance of advantages and disadvantages and whether you think such large-scale schemes are worth developing.

Finally ...

■ Have you answered the questions as set?
■ Have you used all the resources – especially the map and photograph?
■ Have you incorporated data/evidence from the resources?